Modern Computational Techniques for Engineering Applications

Modern Computational Techniques for Engineering Applications presents recent computational techniques used in the advancement of modern grids with the integration of non-conventional energy sources like wind and solar energy. It covers data analytics tools for smart cities, smart towns, and smart computing for sustainable development.

This book

- Discusses the importance of renewable energy source applications in wind turbines and solar panels for electrical grids.
- Presents optimization-based computing techniques like fuzzy logic, neural networks, and genetic algorithms that enhance the computational speed.
- Showcases cloud computing tools and methodologies such as cybersecurity testbeds and data security for better accuracy of data.
- Covers novel concepts on artificial neural networks, fuzzy systems, machine learning, and artificial intelligence techniques.
- Highlights application-based case studies including cloud computing, optimization methods, and the Industrial Internet of Things.

The book comprehensively introduces modern computational techniques, starting from basic tools to highly advanced procedures, and their applications. It further highlights artificial neural networks, fuzzy systems, machine learning, and artificial intelligence techniques and how they form the basis for algorithms. It presents application-based case studies on cloud computing, optimization methods, blockchain technology, fog and edge computing, and the Industrial Internet of Things. It will be a valuable resource for senior undergraduates, graduate students, and academic researchers in diverse fields, including electrical engineering, electronics and communications engineering, and computer engineering.

Modern Computational Techniques for Engineering Applications

Edited by
Krishan Arora
Vikram Kumar
Deepak Prashar
Suman Lata Tripathi

CRC Press
Taylor & Francis Group
Boca Raton London New York

CRC Press is an imprint of the
Taylor & Francis Group, an **informa** business

Front cover image: Jackie Niam/Shutterstock

First edition published 2024
by CRC Press
2385 NW Executive Center Drive, Suite 320, Boca Raton FL 33431

and by CRC Press
4 Park Square, Milton Park, Abingdon, Oxon, OX14 4RN

CRC Press is an imprint of Taylor & Francis Group, LLC

ISBN: 978-1-032-42462-0 (hbk)
ISBN: 978-1-032-52596-9 (pbk)
ISBN: 978-1-003-40740-9 (ebk)

DOI: 10.1201/9781003407409

Typeset in Sabon
by SPi Technologies India Pvt Ltd (Straive)

Contents

Preface

This book presents novel concepts in the development of various computational techniques based on different research areas of Engineering. It discusses the technologies involved in producing efficient and economically feasible modern computational methodologies around the world.

Computational Techniques are a collection of innovative research ideas that provides a complete insight and overview of the applications of modern computational techniques in power and energy. The research focused on a wide range of topics such as artificial neural networks, smart electrical grids, soft computing, and Internet of Things. Computational Methods in Engineering encourages a well-rounded understanding of the subject.

This book is an introduction to Modern Computational Techniques, arranged from basic computing tools to highly developed computational procedures and applications. The main focus is on artificial neural network, fuzzy system, machine learning, and artificial intelligent techniques and how they form the basis for algorithms.

MATLAB® is a registered trademark of The MathWorks, Inc. For product information,

please contact:
The MathWorks, Inc.
3 Apple Hill Drive
Natick, MA 01760-2098 USA
Tel: 508-647-7000
Fax: 508-647-7001
E-mail: info@mathworks.com
Web: www.mathworks.com

About the Editors

Krishan Arora has completed his Ph.D. in Electrical Engineering from MM (Deemed to be) University, Ambala. He did his M.Tech in Electrical Engineering from IKG Punjab Technical University, Punjab, and B.Tech in Electrical and Electronics Engineering from IKG Punjab Technical University, Punjab. He is associated with Lovely Professional University as an Associate Professor with more than 15years of experience in academics. He is currently the Head of Department, Power Systems in School of Electronics and Electrical Engineering, Lovely Professional University from February2017. He has published more than 55 research papers in refereed IEEE, Springer, and IOP Science Journals and Conferences. He has organized several workshops, summer internships, and expert lectures for students. He has supervised more than 10 postgraduate thesis and more than 15 undergraduate student's projects. He has taken and completed 15 Non-Government and Consultancy Projects. He has attended/participated in 28 National/International Online webinars. His areas of expertise include Electrical Machines, Non-Conventional Energy Sources, Load Frequency Control, Automatic Generation Control, and Modernization of Smart Grids. He has taught various courses at the UG and PG level such as Power Electronics, Non-Conventional Energy Sources, Electric Drives, Induction and Synchronous Machines, and Digital Electronics.

Vikram Kumar is currently working as an Associate Professor in the Department of Electrical Engineering at Lovely Professional University, India. He had completed his doctorate in 2017 and currently pursuing Post Doctorate Fellowship from the University of Calgary, Alberta, Canada. He has received his Bachelor of Engineering in Instrumentation and Control Engineering from Maharshi Dayanand University, Rohtak, India (2006), Masters of Technology in Power System Engineering from Rajasthan Technical University, Kota, India (2008), and Ph.D. in Electrical Engineering from I.K. Gujral Punjab Technical University, Jalandhar, India (2017) on the topic "Multi-Objective Multi Area Unit Commitment Problem of Electric Power System." He has published 82 research papers in SCI/referred journals, presented my research work in more than two dozen international and national conferences, and recently applied for three patents on Li-Fi technology for electric vehicle smart parking, charging, and fast tag communication. He had supervised 02 Ph.D. Dissertations, 13 master dissertations, and 20 capstone projects during the last five years. He has copyrights for four recently proposed algorithms in multidisciplinary design and optimization problems. He is a member of IEEE and acts as an editorial board member for the International Journal of Advancements in Electronics Engineering (IJAEE), International Journal of Enhanced Research in Science and Engineering and Technology (IJERSET), International Journal of Electrical and Computer Engineering (IJECE), Senior Member of Universal Association of Computer and Electronics Engineers (UACEE),International Journal of Advancements in Electronics and Electrical Engineering (IJAEEE), International Conference on Mechanics & Applied Physics (ICMAPH' 2015) and Reviewer for Renewable & Sustainable Energy Reviews (Elsevier), Electric Power Components and Systems (Taylor & Francis Group), ADMMET'2015 (EDAS), Energy Systems, IEEE Transactions on Power Systems, IEEE Transactions on Control Systems Technology, IEEE Transactions on Smart Grid, IET Generation, Transmission and Distribution. He had qualified GATE-Instrumentation from IIT Roorkee and got a scholarship from MHRD, Govt. of India, to pursue a Master of Technology. I had got International Travel support from I.K. Gujral Punjab Technical University, India, to present my research work in USA in 2013. I have got an International Travel grant from the Department of Science and Technology, Govt. of India, to attend the International Conference in Canada and Germany during 2016 and 2017. He had efficiently adopted the existing algorithms for the solution of highly constrained power system optimization problems viz. Economic Load Dispatch, Generation Scheduling, Unit Commitment, Multi-Objective Economic load dispatch, and generation scheduling and multi-area unit commitment and modified them to form hybrid algorithms. I had also designed a couple of new meta-heuristics/heuristics and evolutionary search algorithms for engineering optimization and applied them for standard benchmark test systems and the solution of nonconvex economic load dispatch problem, unit commitment problem for single and multi-objective and multi-criteria framework. I had recently developed Intensify Harris Hawks optimizer, hybrid GWO-RES, Hybrid GWO-PS, Hybrid SMA-SA, hybrid SMA-PS, Hybrid SMA-RES, Modified Chimp Algorithm, Chaotic Harris Hawks optimize, Ameliorated Dragonfly Algorithm (ADFA), Hybrid HSRS, Hybrid DE-RS and Hybrid DE-HS algorithm for multidisciplinary design and optimization problems and applied Grey Wolf optimizer, Biogeography Based Optimisation, Ant Lion optimizer, Dragonfly optimizer algorithm, Artificial Neural Network/Closed Loop Control Algorithm/Ions Motion optimizer algorithm, Multi-Verse optimizer, Differential Evolution algorithm, Harmony Search Algorithm, Particle Swarm Optimisation, Genetic Algorithm, Quadratic Programming, Mixed-Integer programming, Hybrid Differential, Evolution-Harmony Search algorithm, Hybrid Differential Evolution-Random Search algorithm, Hybrid Harmony Search-Random Search algorithm for the solution of highly constrained power system optimization problems. He had also proposed an investigation for the association between Govt. Sector and private sector for implementing restricting environment for pool operations of electric utilities for the conservation of energy resources.

His primary research efforts focus on Multidisciplinary Design and Optimisation, Artificial Intelligence and machine learning for Numerical and Engineering Optimisation, ICT for Smart Grid Applications, Smart Grid for V2G and G2V Applications, ICT for Renewable Energy, ICT for Micro grids, AI for Renewable Energy, AI for Medical Field and Wireless Body Area Network, Nature-Inspired Computation and Machine Learning for medical and signal processing applications, Renewable Energy Risk Management and Grid Optimisation, Economic Load Dispatch for quadratic and cubical cost function, Single and Multi-objective Economic Load Dispatch, Single and Multi-objective Economic Load Dispatch including valve point effect, Economic Load Dispatch incorporating wind Power, Economic Load Dispatch incorporating Solar Power, Hydro-Thermal and Wind-Thermal Scheduling of electric power system, Thermal Scheduling incorporating Smart Grids, Hydro-Thermal Scheduling incorporating Smart Grids, Single and Multi-Objective Unit Commitment Problem formulation, Single Objective Unit Commitment incorporating wind power, Single Objective Unit Commitment incorporating solar power, Single and Multi-Objective Unit Commitment Incorporating Smart Grids, Multi-Objective and Multi-Area Unit Commitment Problem, Multi-Objective and Multi-Area Unit Commitment Problem incorporating smart grids, Power System Operation and Control, Vehicle Number Plate Recognition, Biometric and Face Recognition, Pattern Recognition, Automatic Generation Control and Reliability Assessment of Power System.

Deepak Prashar received B.Tech in Computer Science and Engineering from Punjab Technical University, Punjab, India, in 2007 and M.Tech in Computer Science and Engineering from PEC University Of Technology, Chandigarh, India, in 2009. Presently, he is working as an Associate Professor and Head of Department of Networking & Security in Computer Science Department at Lovely Professional University, Punjab, India, since 2009. He has published more than 50 research papers in reputed national, international conferences, and journals, including SCOPUS and SCIE indexed journals. His research interest includes Wireless Sensor Networks, Soft Computing, Block chain, Machine Learning, IoT, Image Processing, and cybersecurity. He is involved in the review process of many SCI or SCIE indexed journals like IEEE, Springer, Wiley, Inderscience, etc. He is also the Editorial Board member of recognized journals and serves as a technical program committee member in reputed international conferences in India and abroad. He is also the Brand Ambassador of Bentham Science Series.

Suman Lata Tripathi has completed her Ph.D. in microelectronics and VLSI from MNNIT, Allahabad. She did her M.Tech in Electronics Engineering from UP Technical University, Lucknow, and B.Tech in Electrical Engineering from Purvanchal University, Jaunpur. She is associated with Lovely Professional University as a Professor with more than 17 years of experience in academics. She has published more than 65 research papers in refereed IEEE, Springer, and IOP science journals and conferences. She has also published 12 Indian patents and two copyrights. She has organized several workshops, summer internships, and expert lectures for students. She has worked as a session chair, conference steering committee member, editorial board member, and peer reviewer in international/national IEEE, Springer, Wiley, etc., Journal and conferences. She has received the "Research Excellence Award" in 2019 and "Research Appreciation Award" in 2020, 2021 at Lovely Professional University, India. She had received the best paper at IEEE ICICS-2018. She has edited and authored more than 14 books/1 Book Series in different areas of Electronics and electrical engineering. She is associated for editing work with top publishers like Elsevier, CRC Taylor &Francis Group, Wiley-IEEE, SP Wiley, Nova Science, Apple academic press, etc. She is also associated as an editor of book Series on "Green Energy: Fundamentals, Concepts, and Applications" and "Design and development of Energy efficient systems" to be published by Scrivener Publishing, Wiley (In production). She is also associated with Wiley-IEEE for her multi-authored (ongoing) book in the area of VLSI design with HDLs. She is also working as a book series editor for the title, "Smart Engineering Systems" and a conference series editor for "Conference Proceedings Series on Intelligent systems for Engineering designs", CRC Press Tylor & Francis Group. She has already completed one book with Elsevier on "Electronic Device and Circuits Design Challenges to Implement Biomedical Applications." She is a guest editor of a special issue in "Current Medical Imaging", Bentham Science. She is associated as a senior member of IEEE, Fellow IETE, and Life member ISC, and continuously involved in different professional activities along with academic work. Her areas of expertise include microelectronics device modeling and characterization, low-power VLSI circuit design, VLSI design of testing, and advance FET design for IoT, Embedded System Design, biomedical applications, etc.

Contributors

Krishan Arora
Lovely Professional University
Phagwara, India

Sandhya Avasthi
ABES Engineering College
Gaziabad, India

Fabián Enrique Casares
University of Pamplona
Colombia

Parul Choudhary
GLA University
Mathura, India

A. Dash
Sambalpur University
Burla, India

N. Dhanyaa
Sri Krishna College of Technology
Coimbatore, India

K. Dinesh
Sri Krishna College of Technology
Coimbatore, India

B. Kamali
Sri Ramakrishna Engineering
   College
Coimbatore, India

K. Kanimozhi
Sri Krishna Adithya College of Arts
   and Science
Coimbatore, India

K. Kathirvel
Sri Krishna College of Technology
Coimbatore, India

Raabia Kausar
Lovely Professional University
Phagwara, Punjab

V. Kaviyanjali
Sri Krishna College of Technology
Coimbatore, India

R. Kiruba
Sri Ramakrishna Engineering
   College
Coimbatore, India

Kakarla Hari Kishore
Koneru Lakshmaiah Education
   Foundation
Guntur, India

P. Rama Krishna
Anurag University
Hyderabad, India

T. Santosh Kumar
CMR Institute of Technology
Hyderabad, India

R. Senthil Kumar
Sri Krishna College of Technology
Coimbatore, India

Daniel Steven Moran
University of Pamplona
Colombia

Maria C. Moreno
University of Pamplona
Colombia

Pooja Pathak
GLA University
Mathura, India

Yogeta Pimpale
Lovely Professional University
Phagwara, Punjab

P. Lenin Pugalhanthi
Sri Krishna College of Technology
Coimbatore, India

V. Radhika
Sri Ramakrishna Engineering
    College
Coimbatore, India

Rukia Rahman
University of Kashmir
Srinagar, India

Umesh C. Rathore
Government Hydro Engineering
    College Bandla, Bilaspur, HP
Bilaspur, India

V. Rukkumani
Sri Ramakrishna Engineering
    College
Coimbatore, India

Brayan Daniel Sarmiento
University of Pamplona
Colombia

P. K. Sethy
Sambalpur University
Burla, India

Dalwinder Singh
Lovely Professional University
Phagwara, India

Janpreet Singh
Lovely Professional University
Phagwara, India

N. Sowndarya
Sri Ramakrishna Engineering
    College
Coimbatore, India

R. Sowndarya
Sri Ramakrishna Engineering
    College
Coimbatore, India

B. Srikanth
Vardhaman College of Engineering
Hyderabad, India

K. Srinivasan
Sri Ramakrishna Engineering
    College
Coimbatore, India

Oscar J. Suarez
University of Pamplona
Colombia

Pradeep Singh Thakur
Rajiv Gandhi Government
    Engineering College (RGGEC)
Masal, India

Suman Lata Tripathi
Lovely Professional university
Phagwara, Punjab

Chapter 1

Paralysis support system using IoT

V. Radhika, B. Kamali, N. Sowndarya and R. Sowndarya
Sri Ramakrishna Engineering College, Coimbatore, India

CONTENTS

1.1 INTRODUCTION

The inability to move muscles independently and purposefully is referred to as paralysis. It could be either temporary or permanent. Stroke, spinal cord injury, and multiple sclerosis are the most common causes. Paresis is a significant weakness that causes a complete loss of movement. Damage to the nervous system, particularly the spinal cord, is the most common cause of paralysis. A central nervous system (brain and spinal cord) injury or disease disrupts nerve signals to the muscles, resulting in paralysis.

Despite new approaches to healing or treating paralyzed patients, the aim of this research is to help paralyzed patients adjust life by becoming as independent as possible. There are large and expensive machines. They appear to be available only in hospitals and cannot be used at home or at patient's leisure. The aim of this work is to develop a device that can retrain patients' movements. The proposed device can be used by patients themselves, is cheap, and can easily interact with others and help patients in an emergency.

DOI: 10.1201/9781003407409-1

1.2 LITERATURE REVIEW

Since the term "Internet of Things"(IoT) was coined, there has been an increase in work and research on the subject. One of the most important steps in the healthcare industry is the integration of IoT devices into systems. It provides an opportunity to eliminate human response delays by having a machine that takes actions in the event of an emergency, like providing medicine for allergy or insulin. There are many related works, some of which are listed below.

Gupta et al. [1] proposed an IoT-based smart healthcare kit that uses specialised biomedical sensors to collect data such as heart rate, blood pressure, and electrocardiogram. Furthermore, it can send an alert to the physician outlining the patient's current status as well as all relevant information. A second-generation Intel Galileo board was used in the above work that processes the data taken from the sensors and produces an output that is readable and easily understandable by the doctor. This database contains all the information, recordings, and medical history of the patients. The mobile app is used to monitor the data, and a web portal was created that could connect to the database server and display a live feed of the patient's vitals.

Kiran et al. [2] proposed a Remote Health Care System. In this work, a decision-making engine is used that is intended to transmit the data with power efficiency. This transmits the data more efficiently. Only one medical sensor is used in this system that collects ECG data. Because of the above proposed idea, this model outperformed other systems using burst transmission.

Chiuchisan et al. [3] propose a monitoring system for high-risk patients at intensive care units (ICU). It monitors the vital signs of the patients and alerts the nurse or caretaker team about the unusual behavior of the patients. This method uses sensors, such as the XBOX KinectTM, to recognize the facial changes and detects even small changes in the patient's movement.

Usage of smart wearables and fitness band is rapidly growing as an alternative to the IoT-based healthcare monitoring systems. The increase in the use of wearable devices is tremendous and improves the personal healthcare monitoring system. The heart rate and oxygen levels are continuously monitored for the patients who are admitted in the hospital. All of the readings are readily available to the patients in their hands. These products are manufactured by the leading manufactures at a low cost. Even though the status of patients is readily available in smart bands and also it can be transferred to their personal mobile phones with special features, these data are not readily available to medical authorities, as these data cannot be accessed from another device [4, 5]. Moreover, these data cannot be stored in a local device or remotely, thus making them ineffective for the intended purpose.

1.3 PROPOSED HEALTHCARE SYSTEM

The proposed work is to create a healthcare system for paralysis patients that will allow them to communicate with doctors, nurses, and family members. To achieve this functionality, the system employs microcontroller-based circuitry. It employs a hand gesture recognition unit as well as a pulse receiver and transmitter circuit. The hand gesture recognition unit detects hand movement and the data are transmitted to the receiver system. The commands are processed by the receiver system and displayed on the LCD. The block diagram of the proposed unit is shown in Figure 1.1.

GY-521 accelerometer, NodeMCU (ESP-32), HT12E encoder, and RF433 transmitter module comprise the transmitter section. A NodeMCU (ESP-32), an HT12D decoder, and an RF433 receiver module comprise the receiver section. The cost of NodeMCU is low, consumes less power with Wi-Fi, and has an in-built dual-mode Bluetooth. The GY-521 unit is a breakout board for the MEMS (Micro electromechanical systems), which have a three-axis gyroscope, an accelerometer, a motion processor, and a temperature sensor [6].

An accelerometer is linked to the ESP32 in the transmitter section. When the direction of the accelerometer changes, the initial or stable value of the

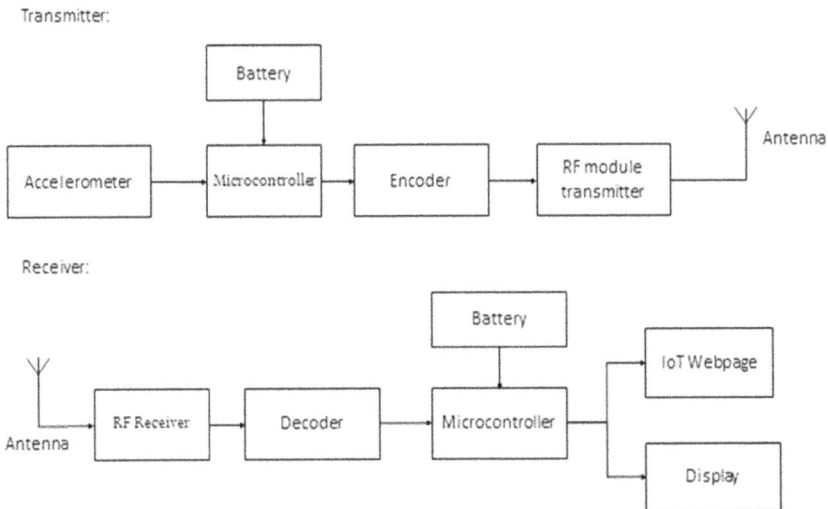

Figure 1.1 Block diagram of proposed paralysis support system.

accelerometer changes. The primary unit is an accelerometer capable of detecting gestures caused by changes in the position. The sensor provides an analogue variation of voltage on its x, y, and z pins in response to the position change. The analogue variation is converted into digital format by using an Op-Amp as a comparator in which a threshold voltage (comparison) is preset and the voltage is either high or low depending on the input voltage [7]. The Op-Amp IC LM324 is used for this application. This unit produces various four-bit binary sequences upon change of its position. This sequence is transmitted through the RF channel. The transmitter unit attached to the accelerometer circuit will produce parallel binary data as an input to the HT12E encoder. It encodes the generated parallel data to serial data. This serial data is transmitted over a 434-MHz carrier channel with ASK modulation through a short dipole antenna. Accelerometer-embedded transmitter phase is installed on a movable component of the patient's body. Patients want to make simple gestures to express their wishes [8]. Messages are displayed on the LCD in response to the gestures [9].

1.4 SYSTEM ARCHITECTURE

The system architecture for this work has two main blocks that work together for hand gesture recognition, the collected data is mapped into its corresponding sign and sign is converted into alphabets [10]. The hardware setup for the proposed work is shown in Figure 1.2.

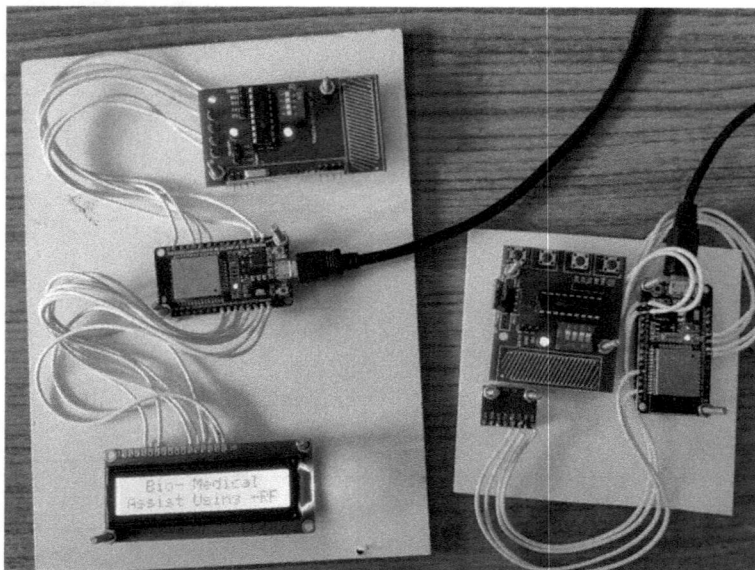

Figure 1.2 Hardware setup of paralysis support system.

1.4.1 ESP32-NODE MCU

The ESP32 [11] microcontroller series consumes less power and is available at a low cost. It is a system-on-a-chip microcontroller with integrated Wi-Fi and dual-mode Bluetooth. The ESP32 series was powered by a dual-core or single-core RISC-V microprocessor and have default antenna switches, an RF balun, an amplifier, filters, and power-management units. The Node MCU was manufactured by a Chinese company using 40 nm process.

1.4.2 Gyro and accelerometer sensor

Gyro sensors are angular velocity sensors, also called as angular rate or velocity sensors. The most common unit of measurement for angular velocity is degrees per second. The acceleration or rate of change in velocity of the human body in its instantaneous rest frame is measured by an accelerometer. This is distinct from coordinate acceleration, which refers to acceleration in a fixed coordinate system. The GY-521 unit is a breakout board for the MEMS, which has a three-axis gyroscope, an accelerometer, a motion processor, and a temperature sensor [6]. The complex algorithms can be easily processed by the motion processor on the board. The motion processor converts the raw sensor values into stable position data by processing the algorithms. I2C serial data bus is used to read the sensor values and it requires two wires (SCL and SDA).

1.4.3 RF module

This module, as the name suggests, operates at radiofrequency (RF). The range of RF lasts between 30 kHz and 300 GHz. In this type of system, variation in amplitude of carrier wave represents the digital data. This type of modulation can be called as Amplitude Shift Keying. For a variety of reasons, RF transmission outperforms IR transmission (infrared). To begin with, these signals can be used for long-distance applications and primarily the infrared operates on in line-of-sight mode; these signals can travel even when there is an obstacle between the transmitter and the receiver. Following that, RF transmission is more powerful and dependable than IR transmission. An RF transmitter and a receiver are included in this module and they operate at 434 MHz. The transmitter receives serial data and transmits wirelessly over RF with the help of an antenna connected to the pin4, and the RF receiver also operates at the same frequency as the transmitter. This module is frequently used in encoders/decoders. Encoding of parallel data for transmission is done by the encoder, and the decoder is responsible for decoding the received data.

1.4.4 Encoder and decoder

The HT12E Encoder ICs are a CMOS LSI family intended for use in remote control systems. They can encode a total of 12 bits of data, including address of eight bits and the data of four bits. The address and data input can be externally programmed or fed through the switches.

The HT12D Decoder ICs are a CMOS LSI family intended for use in remote control systems. After processing the data, the decoder receives the address and data transmitted by a carrier wave through the RF transmission medium and provides output to the output pins.

1.4.5 LCD display

The display technology used in notebook computers and other small computers is LCD (liquid crystal display). The passive or active matrix forms the LCD display grid. The light of any pixel can be controlled by sending the current across two-grid conductors.

1.4.6 Arduino IDE

The software used for the development of this work is Arduino IDE that is very light and easy to use.

It is a free and open source software for writing and compiling the code for the Arduino Module. It is software that makes the compilation of code so simple and easy for beginners to learn. It is readily available for almost for all the operating systems and runs on the Java Platform that includes functions and commands to debug, edit, and compile the code in the environment. The microcontroller on the board can be programmed and accepts information in the form of code. The main code can also be called as sketch; writing on the IDE platform generates a Hex File, which can be transferred and uploaded into the controller of the board. The IDE environment is primarily comprised of two basic components: the Editor and the Compiler; the editor is used for writing the code and the compiler is used for compiling and uploading the program into the appropriate Arduino Module. This is compatible with C and C++ programming languages.

1.5 RESULTS AND DISCUSSION

Results include the successful operation of the paralysis patient support system. An accelerometer and a gyro sensor along with RF transmitter are placed in the moving part of the patient. The receiver part along with an LCD is placed near the caretaker. Whenever the patient needs something, they can make some movement and it will be detected by the accelerometer sensor and the signal will be sent to the receiver. The receiver will receive the signal and it will be displayed in the LCD. The system helps the paralysis patient to convey their needs to the caretaker. The LCD display result is shown in Figure 1.3.

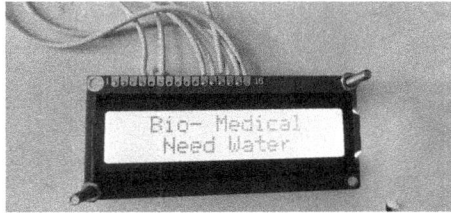

Figure 1.3 Result displayed in LCD.

1.6 CONCLUSION AND FUTURE SCOPE

Although there are several systems in place to monitor the health of paralyzed patients, there are few systems that focus on communication. However, this system bridges the communication gap between the patient and others, allowing the paralyzed patient to relieve stress by revealing their thoughts and motivating them as much as possible. And this system is inexpensive without incurring significant debt, while also being functional.

For the future work, the number of health sensors can increase and hence help the patient further. If a patient is not attended to for an extended period of time and requires water or food, an alert message can be sent to the patient's mobile phone. With the help of this system, relatives and doctors can keep a better eye on patients. The system can be made smarter and more efficient in the future by making the goggle wireless for eyeblink detection. This is possible using Bluetooth and Wi-Fi technology, so that the system is efficient, secure, and simple to use. Additionally, for continuous patient monitoring, security indicators such as light indicators can be added. A graphical LCD can show a graph of the rate of change of health parameters over time. The entire patient monitoring system can be integrated into a small compact unit, which is the size of a cell phone or a wrist or a smart watch. This device is simple to use for patients or other people.

REFERENCES

1. P. Gupta, D. Agrawal, J. Chhabra, P. K. Dhir (2016), "IoT based smart health-care kit", *Proceedings of the Computational Techniques in Information and Communication Technologies (ICCTICT)*, pp. 237–242.
2. M. P. R. S. Kiran, P. Rajalakshmi, K. Bharadwaj, A. Acharyya (2014), "Adaptive rule engine based IoT enabled remote health care data acquisition and smart transmission system", *Proceedings of IEEE World Forum on Internet of Things*, pp. 253–258.
3. I. J. Kosmas, T. Papadopoulus, C. Michalakelis (2021), "Applying internet of things (IoT) in healthcare: a survey", *Crimson Publishers*, vol. 4, no.4, pp. 323–329.

4. E. N. Ganesh (2019), "Health monitoring system using Raspberry Pi and IOT", *Oriental Journal of Computer Science and Technology*, vol. 12, no. 1, pp. 392–399.
5. S. Banka, I. Madan, S. S. Saranya (2018), "Smart healthcare monitoring using IOT", *International Journal of Applied Engineering Research*, vol. 13, no. 15, pp. 189–192.
6. J. Lin, Y. Chang, C. Liu, K. Pan (2011), "Wireless sensor networks and their applications to the healthcare and precision agriculture, wireless sensor networks", *International Journal of Engineering Trends and Technology*, vol. 1, no. 1, 342 p.
7. A. Botre (2016), "Assistance system for paralyzed", *International Journal of Innovative Research in Electrical, Electronics, Instrumentation and Control Engineering*, vol. 4, no. 5, pp. 89–98.
8. T. Deepasri, M. Gokulpriya, G. Arun Kumar, P. Mohanraj, M. Shenbagapriya (2016), "Automated paralysis patient health care monitoring system", *South Asian Journal of Engineering and Technology*, vol. 3, no. 2, pp. 85–92.
9. M. Kumar, K. Pandurangan, R. Vinu (2021), "Automated paralysis patient monitoring system", *Proceedings of IEEE National Biomedical Engineering Conference (NBEC)*, pp. 71–76.
10. V. J. Handan, N. K. Choudhari (2020), "Automated paralysis patient health care system", *International Journal of Infinite Innovations in Technology*, vol. 6, no. 3, pp. 01–03.
11. K. Arora (2022), "Internet of Things-based modernization of smart electrical grid", *Smart Electrical Grid System*, USA: CRC Press.

Chapter 2

Blockchain and its applications

A review

Parul Choudhary and Pooja Pathak
GLA University, Mathura, Uttar Pradesh

CONTENTS

2.1 INTRODUCTION

Unlike conventional approaches, blockchain allows digital assets without intermediaries to be transferred peer-to-peer [1]. Blockchain was initially developed to support the popular Bitcoin cryptocurrency. Bitcoin was first suggested by Nakamoto in 2008 and put into effect in 2009 [2]. In 2016, the stock market grew by $10 billion. Blockchain is a chain that records all transactions carried out through a public directory [3]. When new blocks are attached to it, the chain grows constantly. Decentralized blockchain is composed of various key technologies such as digital signatures, hash encryption, and distributed consensus algorithms. All the transactions are carried out in a decentralized way eliminating the duty to confirm and check transactions through any intermediary [4]. The characteristics of blockchain include decentralization, accountability, fixity,and suitability [5].

Even though bitcoin is the most common blockchain technology, it has applications beyond cryptocurrencies. Meanwhile, it facilitates transactions without a bank or broker, and can be used for various financial services like digital assets, transfer services, and payment online services [6]. Bitcoin itself has taken own lives and permeated numerous industries like banking, healthcare, government, produce and supply, with a huge number of applications [7]. The blockchain can evolve and transform into a broad spectrum of applications, such as in product transfers (source chain), digital broadcasting transfers (selling art), remote services (tourism and travel), networks such as data transfers to computer sources, and distribute credentials. Blockchain also provides distributed tools (production and supply of electricity), crowdfunding, e-voting, identity management, and government documents.

Blockchain is an all-cryptocurrency exchange digitalized, decentralized and accessible booklet. The transactions are chronologically recorded and assist participants in keeping track of transactions with digital currencies without central logging [8–10]. The distributed leader is one of the big features of blockchain [11]. In several copies through computers forming a peer-to-peer network, this sort of record denotes the absence of a single, centralized database or server [12]. Instead, there is a decentralized computer network-wide blockchain database. Each network machine is called a network node and each network node is provided with a spare copy of block shaft that is downloaded automatically. Transactions are digitally signed using two keys–private key and public key. Both means refer to each other mathematically. Because of the sophistication of mathematics, these keys can hardly be guessed, making them harder to break transactions. The public key signs and encrypts messages, and your private password decrypts them. For all new transactions, the blockchain database is distributed to any node as a "World Wide Ledger."

2.2 BLOCKCHAIN ARCHITECTURE

Blockchain is expected to have a big impact on all industries soon. Financial institutions are ingeniously designing ways to evaluate and invest in this technology making an understanding of the structure and the working algorithm of the blockchain technology highly relevant for everyone. A blockchain is a collection of records called cryptographically connected and secured blocks. Every block contains a cryptographic timestamp, hash and transaction data. This data structure is powerful, and the adjacent transaction blocks list can be stored in modest databases or files. These blocks are linked with the previous block in the chain being referred to each other. The first element of the chain is known as "block of genesis." The ledger is known to be upright stack and the blocks stack each other and the block of genesis is the foundation of the stack [13, 14].

If a block has one parent at any given stage, it may have multiple children temporarily. The same block is used for each child in the chain and the same hash values for the parent. While the situation of multiple children primarily comes when a blockchain gap is found, the blocks in gap will leave and not followed in future till these gaps are solved and the valid block is identified.

The child blocks are similar and thus vary according to the identity of the parent block. This results in shifting the preceding block hash point of the child block. The loop continues until the blocks hit the grandchild. The cascade effects ensure that after several generations in a block, all blocks cannot be interfered with or without a convincing recalculation. Figure 2.1 illustrates the layout of the blockchain as follows:

a. **Data:** Depending on the method, the data is saved in the blockchain. It can be applied to a framework of peer-to-peer files like IPFS, distributed databases like Apache Cassandra, Storj, Sia and Ethereum, and so on. The stored data can be used for various purpose such as registration of transaction data, accounting, contracting and IoT.

b. **Hash:** It takes any length input and produces one fixed output. Changing one input value drastically changes its output. Hash functions of blockchain technology are used everywhere. Each data block is hashed and there may be significant or small changes. Alex, a user named for example, attempts to alter the data in block. The changed block then has an entirely different hash value, ensuring any node or miner in the network is aware of the update by updating the ledger copy of all users. This increases the trust of the stored data of the blockchain. Each node is shown in a hash/Merkle tree as a leaf and marked with a block. With this tree, the user can save large data structures safely and efficiently.

c. **Timestamping:** The time when the block was produced must be registered. A timestamp is a tool used to securely monitor the time a record is produced or updated. A process is now an important tool for companies, since it allows the parties to identify a paper in a given moment.

d. **nBits:** Present portable malware target.

e. **Nonce:** Its value is simply a 4-byte value, which starts from 0 and increases every time hash is determined.

2.3 WORKING OF BLOCKCHAIN

Blockchain is a public directory of several processes and blockchain function comprises numerous processes discussed in this regard:

1. The node or user who wish to start transaction records and transmits the data to network.
2. The node or user receiving the data tests in the network for validity. The reviewed data is then saved in a block.
3. Any node or user of network confirms the transaction, either by performing its working algorithm or the stake algorithm evidence in a component needed for authentication.

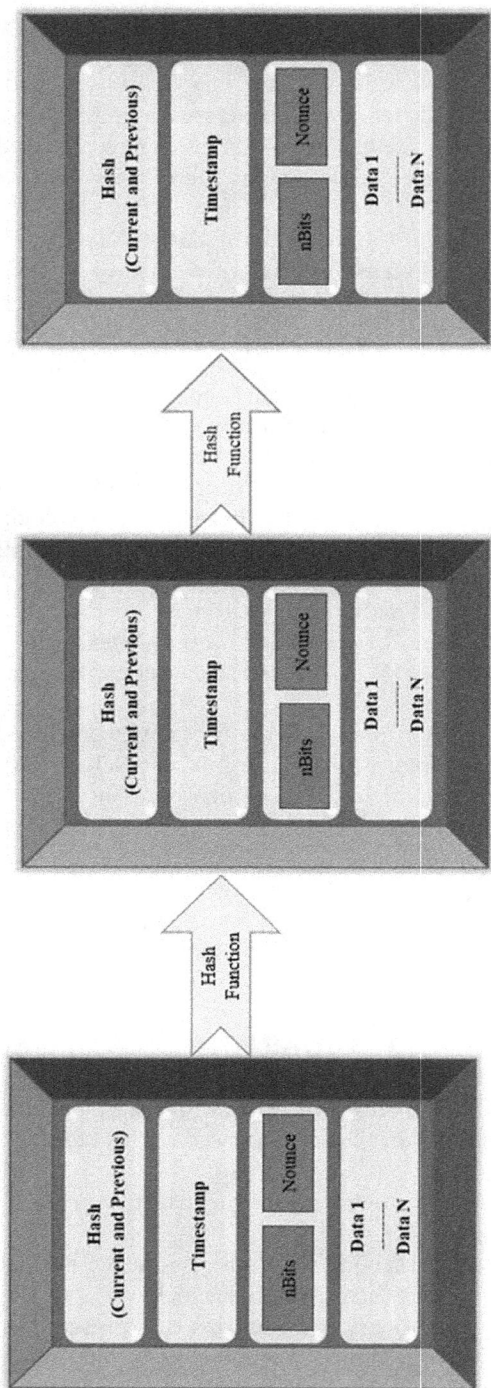

Figure 2.1 Architecture of blockchain.

4. The network's consensus algorithm will store data in block added to blockchain and all network nodes admit the block and stretch the base of the chain on the block.

First, identify the sender identity, indicating that the source and nobody else requests the transaction between the recipient and the source. Figure 2.2 illustrates the checking method using a basic example of a Bob-Alice transaction. Let us say Bitcoin balances are with Alice and Bob, and Alice is willing to pay Bob 10 Bitcoins. Now for sending the money, Alice sends an information message to the blockchain network. Blockchain uses digital signatures to do this (public and private keys) [15]. In addition to its public key, and digital signature, Alice will provide Bobs with details for broadcast, such as its public address and transaction number. To make this digital signature, Alice used her private key. All miners perform autonomous validation of the transaction based on the various criteria we addressed later in this section. It is used as an algorithm of elliptical curve digital signature (ECDSA) [16]. This process makes sure that only their actual owners can invest the funds.

2.4 APPLICATION

In this section, various uses of blockchain technologies are widely discussed. Furthermore, several groups are categorized into applications such as education, health and insurance, and other novel ones.

2.4.1 Healthcare

In today's healthcare systems, blockchain has a huge potential to address issues of interoperability [17]. This standard can be used for safe sharing of Electronic Health Registers (EHR) with stakeholders, such as health departments, physicians, etc. [18]. Sharing of EHR helps us to boost patient quality [19] and to strengthen, for example, the recommendation for physicians [16]. However, it is not a simple task to handle health information, that is, obtain, store, and analyze, particularly for privacy issues. Health data should not be shared with parties who may use it fraudulently.

A health data gateway (HDG), built in the blockchain storage platform, is planned to better address these concerns by [17]. It is a mobile application that can easily handle and monitor the sharing of data. The framework proposed helps users to process medical information without disclosing the privacy of the patient. A private blockchain cloud is often used for storing data, but no one, even doctors and patients, can change the medical data.

The work [18] highlights the design of a modern patient agency framework known as MedRec. It is a spread protocol for blockchain development using public-key cryptography. On each node in the network, blockchain

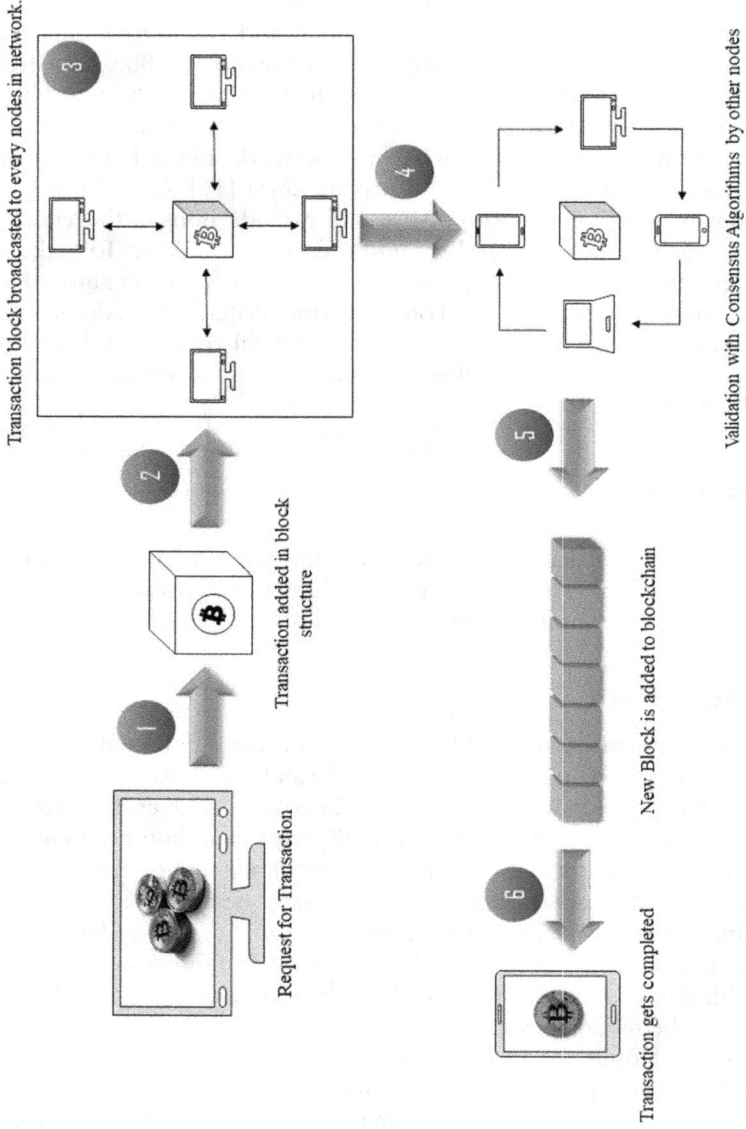

Figure 2.2 Working of blockchain.

replicas are distributed. Like previous work, blockchain is used to monitor access for the automation and monitoring of certain tasks, for example, adding a new record, changing viewership rights, and so on. Also, intelligent Ethereum blockchain contracts [19] are used for constructing an intelligent EHR representation stored in any individual node.

Subsequently [20], PSN is suggested to use blockchain to apply a PSN (a popular network of social networks) to the healthcare sector. PSN healthcare comprises two primary security protocols –a protocol for authenticating and sharing of block chaining data in the PSN region, that is, a protocol in the wireless body network between medical sensors and mobile devices (WBAN). Each PSN node generates and transmits medical data (node address and health sensors). The miners, however, are responsible for verifying transactions and generating new blocks.

Finally, a control system for blockchain access is suggested by [21]. Control of access requires process recognition, authentication, and authorization. This specifies that a user's access to which specific activity on a device can be traced is accountable. This framework lets users access EHR in shared data pools using blockchain after verifying its identity and cryptographic keys. Authentication based on identification is performed to authenticate the individual. Moreover, the latest blockchain implementation can be improved with an optimized lightweight block format.

2.4.2 Education

Blockchain networks were proposed to help keep the education process immutable. Some proposals document innovative work or ideas for creating a scholarly reputation [22]; that constantly report the activities of the learner within different learning organizations [23]; that enable worldwide higher schools to award credit for their courses [24]; that allow for certificates of education [25]; and other records to be given and revoked [26].

The reporting of wider learning activities like voluntary service is a related subject. Salah et al. [3] define a life-long voluntary blockchain scheme. Contrary to other scheduling and task allocation schemes, it fosters an open market for voluntary work, which encourages intelligent coordination, gaming, and personal development. Blockchain stores the permanent digital signatures for voluntary work, reviews, and qualifications and also gives volunteers the sovereignty of data. [27] is a framework that explores how to monitor time and activity information for volunteers by using blockchain.

2.4.3 Insurance

Enhancing the insurance market can also be helpful for individuals. Xu et al. [28] addressed blockchain application to the life cycle of insurance, the request for a binding offer to a policy contract, and the claim procedure, which may lead to the reduction of false insurance claims. Experimental prototypes to provide insurance policy control have been developed [29].

Specific insurance topics are attracting greater consideration. In an insurance-as-you-go car insurance scheme, a case of the use of micro insurance has been introduced [30] to monitor and analyze data. It makes drivers who seldom use cars only for those travels that they choose to travel insurance premiums. A prototype for fake insurance on a blockchain basis was developed [31]. The framework aims to provide automated input on the parties involved in real-time and immutably, thus offering stable distributed infrastructure for cyber risk assessment.

2.4.4 E-Commerce

An E-Commerce Platform based on peer-to-peer blockchains [32] is reported in the trade market and used by workers of a major multinational corporation. The efforts of blockchain to tackle fabricated products on the market are also used. Raikwar et al. [33] explain how goods in blockchain are registered as ownership. To control product ownership, a blockchain prototype has been deployed for post supply chain to identify fake goods by customers [34].

Blockchain's data integrity feature used for other online business determinations, in particular [35] to avoid fraudulent service commission, forged digital ads. The framework allows the user to connect several publicity reports in a way similar to blockchain architecture. In addition to built-in social behavior trends, this blockchain enables marketers to classify authentic ad reports.

In addition to online business, automated physical sales systems like sellers may use technology for blockchain. Automated distribution systems record the quantity of the product and the details of sales in blockchain [36], allowing users to always obtain the current system's product information. An association between humans and machines is recorded in hybrid transactions [37]. This suggests the theoretical plan of a blockchain framework to ensure that messages controlling an intelligent door lock are central and unrepudiated. Intelligent door closure tests validity controls and records the blockchain door checks [38] is another blockchain project of transaction application, which involves computer and robot hybrid financial transactions.

2.4.5 Transportation

For the broad car markets, information on the automotive life cycle is important. Mamais and Theodorakopoulos [39] introduced a blockchain-based vehicle facts registration and administration framework to recover transparency, odometer, and other fraud. There is a broad debate on blockchain in the transport of goods to digitize shipping documents, the loading letter, and compliance exchanges [40, 41], which are potentially cost-cutting in international trade [42]. offers a custom blockchain implementation that facilitates tamper-saving data traceability and automates the monitoring of regulatory conformity [43]. Provides a different blockchain prototype to track cargo and handle various supply chain tasks.

2.4.6 Industry of strength

The power sector has undergone major changes in recent years by incorporating emerging technologies and new energy sources. Due to the mass of mobile, IOT-enabled devices (smartphones, smart meters, and electric vehicles), varying energy needs and energy networks are becoming increasingly complex. Blockchain can accelerate this global power transition by cutting transaction costs and operating the grid more effectively as a platform [44]. The smart agreements between the various smart grid components and devices allow for optimized grid operations. In addition, intelligent contracts on the basis of a network like Ethereum allow blockchain to encourage buyers (both consuming and producing) by enabling it to monetize its abundant power by securely collecting and sending and receiving instalments [45–50].

2.5 CONCLUSION

In this chapter, we provide insights into the value of blockchain technology for different intelligent applications in which protection is still essential. This study is split into four parts. The first section addresses the blockchains' history, safety, privacy, honesty, and background. The second section explains the basic architecture of the blockchain. The third part includes the verification of each distributed network transaction that makes the Bitcoin Transaction and Ethereum Transaction Data and Information Ledger permanent, validated, and unalterable article. Our paper focuses on real-time blockchain applications in healthcare, education, insurance, electrical trade, transport, and energy. We continue to conduct a comprehensive study into the future of blockchains.

REFERENCES

1. Aste, Tomaso, Paolo Tasca, and Tiziana Di Matteo. "Blockchain technologies: The foreseeable impact on society and industry." *Computer* 50.9 (2017): 18–28.
2. Nakamoto, Satoshi. *Bitcoin: A peer-to-peer electronic cash system.* Manubot, 2019.
3. Salah, Khaled, et al. "Blockchain for AI: Review and open research challenges." *IEEE Access* 7 (2019): 10127–10149.
4. Litke, Antonios, Dimosthenis Anagnostopoulos, and Theodora Varvarigou. "Blockchains for supply chain management: Architectural elements and challenges towards a global scale deployment." *Logistics* 3.1 (2019): 5.

5. Kouhizadeh, Mahtab, and Joseph Sarkis. "Blockchain practices, potentials, and perspectives in greening supply chains." *Sustainability* 10.10 (2018): 3652.
6. Peters, Gareth, Efstathios Panayi, and Ariane Chapelle. "Trends in cryptocurrencies and blockchain technologies: A monetary theory and regulation perspective." *Journal of Financial Perspectives* 3.3 (2015): 67–84.
7. Al-Jaroodi, Jameela, and Nader Mohamed. "Blockchain in industries: A survey." *IEEE Access* 7 (2019): 36500–36515.
8. Casino, Fran, Thomas K. Dasaklis, and Constantinos Patsakis. "A systematic literature review of blockchain-based applications: Current status, classification and open issues." *Telematics and Informatics* 36 (2019): 55–81.
9. Han, Meng, et al. "Generating uncertain networks based on historical network snapshots." *International computing and combinatorics conference*. Springer, Berlin, Heidelberg, 2013.
10. Ji, Shouling, et al. "Whitespace measurement and virtual backbone construction for cognitive radio networks: From the social perspective." *2015 12th annual IEEE international conference on sensing, communication, and networking (SECON)*. IEEE, Korea, 2015.
11. Duan, Zhuojun, et al. "Truthful incentive mechanisms for social cost minimization in mobile crowdsourcing systems." *Sensors* 16.4 (2016): 481.
12. Han, Meng, et al. "Neighborhood-based uncertainty generation in social networks." *Journal of Combinatorial Optimization* 28.3 (2014): 561–576.
13. Glaser, Florian. "Pervasive decentralisation of digital infrastructures: A framework for blockchain enabled system and use case analysis." *Proceedings of the 50th Hawaii international conference on system sciences*. Waikoloa Village, Hawaii, January 4–7, 2017, 2017.
14. O'Dwyer, Karl J., and David Malone. "Bitcoin mining and its energy footprint." *IEEE proceedings conference*. Trento, Italy, 280–285, 2014.
15. Decker, Christian, and Roger Wattenhofer. "Information propagation in the bitcoin network." *IEEE P2P 2013 proceedings*. IEEE, Trento, Italy, 2013.
16. Delgado-Segura, Sergi, et al. "Analysis of the Bitcoin UTXO set." *International conference on financial cryptography and data security*. Springer, Berlin, Heidelberg, 2018.
17. Mettler, Matthias. "Blockchain technology in healthcare: The revolution starts here." *2016 IEEE 18th international conference on e-health networking, applications and services (Healthcom)*. IEEE, USA, 2016.
18. Linn, Laure A., and Martha B. Koo. "Blockchain for health data and its potential use in health it and health care related research." *ONC/NIST use of blockchain for healthcare and research workshop*. ONC/NIST, Gaithersburg, Maryland, 2016.
19. Tama, Bayu Adhi. "Learning to prevent inactive student of Indonesia Open University." *JIPS* 11.2 (2015): 165–172.
20. Tama, Bayu Adhi, and Kyung-Hyune Rhee. "Tree-based classifier ensembles for early detection method of diabetes: An exploratory study." *Artificial Intelligence Review* 51.3 (2019): 355–370.
21. Yue, Xiao, et al. "Healthcare data gateways: Found healthcare intelligence on blockchain with novel privacy risk control." *Journal of Medical Systems* 40.10 (2016): 218.
22. Azaria, Asaph, et al. "Medrec: Using blockchain for medical data access and permission management." *2016 2nd international conference on open and big data (OBD)*. IEEE, Italy, 2016.

23. Zhang, Jie, Nian Xue, and Xin Huang. "A secure system for pervasive social network-based healthcare." *IEEE Access* 4 (2016): 9239–9250.
24. Xia, Qi, et al. "BBDS: Blockchain-based data sharing for electronic medical records in cloud environments." *Information* 8.2 (2017): 44.
25. Sharples, Mike, and John Domingue. "The blockchain and kudos: A distributed system for educational record, reputation and reward." *European conference on technology enhanced learning*. Springer, Cham, 2016.
26. Ocheja, Patrick, Brendan Flanagan, and Hiroaki Ogata. "Connecting decentralized learning records: A blockchain based learning analytics platform." *Proceedings of the 8th international conference on learning analytics and knowledge*. Italy, 2018.
27. Turkanović, Muhamed, et al. "EduCTX: A blockchain-based higher education credit platform." *IEEE Access* 6 (2018): 5112–5127.
28. Xu, Yuqin, et al. "ECBC: A high performance educational certificate blockchain with efficient query." *International colloquium on theoretical aspects of computing*. Springer, Cham, 2017.
29. Chen, Zhixong, and Yixuan Zhu. "Personal archive service system using blockchain technology: Case study, promising and challenging." *2017 IEEE international conference on AI &mobile services (AIMS)*. IEEE, USA, 2017.
30. Kapsammer, Elisabeth, et al. "iVOLUNTEER: A digital ecosystem for life-long volunteering." *Proceedings of the 19th international conference on information integration and web-based applications & services*. USA, 2017.
31. Zhou, Ning, Menghan Wu, and Jianxin Zhou. "Volunteer service time record system based on blockchain technology." *2017 IEEE 2nd advanced information technology, electronic and automation control conference (IAEAC)*. IEEE, China, 2017.
32. Nath, Indranil. "Data exchange platform to fight insurance fraud on blockchain." *2016 IEEE 16th international conference on data mining workshops (ICDMW)*. IEEE Computer Society, USA, 2016.
33. Raikwar, Mayank, et al. "A blockchain framework for insurance processes." *2018 9th IFIP international conference on new technologies, mobility and security (NTMS)*. IEEE, Italy, 2018.
34. Vo, Hoang Tam, et al. "Blockchain-based data management and analytics for micro-insurance applications." *Proceedings of the 2017 ACM on conference on information and knowledge management*. USA, 2017.
35. Lepoint, Tancrede, Gabriela Ciocarlie, and Karim Eldefrawy. "BlockCIS—A blockchain-based cyber insurance system." *2018 IEEE international conference on cloud engineering (IC2E)*. IEEE, USA, 2018.
36. Ying, Wenchi, Suling Jia, and Du Wenyu. "Digital enablement of blockchain: Evidence from HNA group." *International Journal of Information Management* 39 (2018): 1–4.
37. Chang, Po-Yeuan, Min-Shiang Hwang, and Chao-Chen Yang. "A blockchain-based traceable certification system." *International conference on security with intelligent computing and big-data services*. Springer, Cham, 2017.
38. Toyoda, Kentaroh, et al. "A novel blockchain-based product ownership management system (POMS) for anti-counterfeits in the post supply chain." *IEEE Access* 5 (2017): 17465–17477.
39. Mamais, Stylianos S., and George Theodorakopoulos. "Behavioural verification: Preventing report fraud in decentralized advert distribution systems." *Future Internet* 9.4 (2017): 88.

40. Yoo, Minjae, and Yoojae Won. "Study on smart automated sales system with blockchain-based data storage and management." *Advances in computer science and ubiquitous computing.* Springer, Singapore, 2017, 734–740.
41. Han, Donhee, Hongjin Kim, and Juwook Jang. "Blockchain based smart door lock system." *2017 international conference on information and communication technology convergence (ICTC).* IEEE, USA, 2017.
42. Cardenas, Irvin Steve, and Jong Hoon Kim. "Robot-human agreements and financial transactions enabled by a blockchain and smart contracts." *Companion of the 2018 ACM/IEEE international conference on human-robot interaction.* USA, 2018.
43. Brousmiche, Kei Leo, et al. "Digitizing, securing and sharing vehicles life-cycle over a consortium blockchain: Lessons learned." *2018 9th IFIP international conference on new technologies, mobility and security (NTMS).* IEEE, 2018.
44. Jabbar, Karim, and Pernille Bjørn. "Infrastructural grind: Introducing blockchain technology in the shipping domain." *Proceedings of the 2018 ACM conference on supporting groupwork.* IEEE, USA, 2018.
45. Loklindt, Christopher, Marc-Philip Moeller, and Aseem Kinra. "How blockchain could be implemented for exchanging documentation in the shipping industry." *International conference on dynamics in logistics.* Springer, Cham, 2018.
46. Lu, Qinghua, and Xiwei, Xu. "Adaptable blockchain-based systems: A case study for product traceability." *IEEE Software* 34.6 (2017): 21–27.
47. Gao, Zhimin, et al. "Coc: A unified distributed ledger-based supply chain management system." *Journal of Computer Science and Technology* 33.2 (2018): 237–248.
48. Mengelkamp, Esther, et al. "A blockchain-based smart grid: Towards sustainable local energy markets." *Computer Science-Research and Development* 33.1–2 (2018): 207–214.
49. Sanseverino, Eleonora Riva, et al. "The blockchain in microgrids for transacting energy and attributing losses." *2017 IEEE international conference on internet of things (iThings) and IEEE green computing and communications (GreenCom) and IEEE cyber, physical and social computing (CPSCom) and IEEE smart data (SmartData).* IEEE, USA, 2017.
50. Arora, K., A. Kumar, V.K. Kamboj, D. Prashar, B. Shrestha, and G.P. Joshi "Impact of renewable energy sources into multi area multi-source load frequency control of interrelated power system." *Mathematics* 9 (2021): 186.

Chapter 3

Data analytics tools for smart cities and smart towns

K. Kanimozhi

Sri Krishna Adithya College of Arts and Science, Coimbatore

CONTENTS

3.1 INTRODUCTION

The use of sensors, the increased participation of citizens in social networks, and the creation of content are some of the reasons that have contributed to the significant increase in information available in cities [1, 2]. Keeping in mind that cities are interconnected systems, large amounts of data are generated by the government and citizens, and over a period of time, these data are converted into information that can be used to solve problems, reduce uncertainty, and improve the quality of life in cities [3]. Big data enables cities to improve their decision-making processes and their ability to respond to citizen needs through processes that integrate large amounts of data, analytics, and predictions in real time [4].

3.2 THE CONCEPT OF BIG DATA ANALYTICS

Data analytics is the science of analyzing raw data to draw conclusions about that information. Many techniques and processes of data analysis have been automated into machine processes and algorithms that operate on raw data for human consumption [5].

- Data analytics is the science of analyzing raw data to draw conclusions about that information.
- Data analysis helps a business improve performance, operate more efficiently, increase profitability, or make more strategic decisions.
- Techniques and processes of data analytics have been automated into machine processes and algorithms that operate on raw data for human consumption.
- Different approaches to data analysis include what happened (descriptive analysis), why it happened (diagnostic analysis), what is going to happen (predictive analysis), or what needs to be done next (descriptive analysis).
- Data analysis relies on various software tools ranging from spreadsheets, data visualization and reporting tools, data mining programs, or open-source languages.

Figure 3.1 Roles of Big Data analytics.

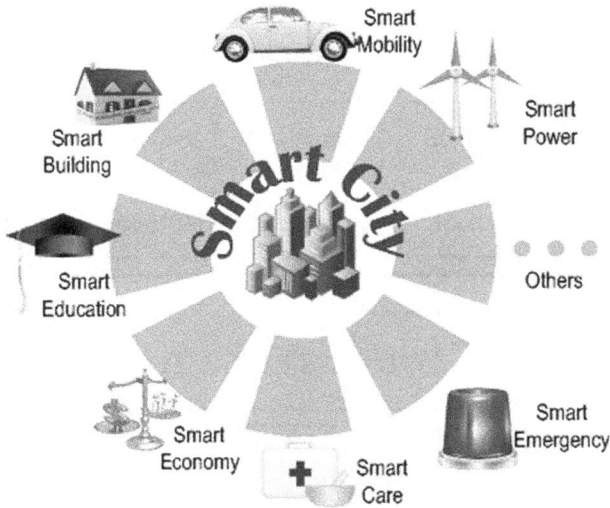

Figure 3.2 Simple demonstration of smart city.

3.3 THE CONCEPT OF SMART CITIES

Smart cities use sensors and connected gadgets to collect and evaluate data. This information is used to improve city operations, manage resources, and improve the daily lives of citizens. Smart cities use technology to improve public transport, traffic management, water and electricity, law enforcement, schools, and hospitals.

3.3.1 Data storage in smart cities

Smart cities can store their data in three main locations:

- **Cloud Storage:** For analytical reasons, smart cities require vast amounts of data. In their data centers, cloud data systems use solid state drives, eliminating redundant data and securing data transmission. Payment options for cloud-based systems are often more flexible than those for on-premise data centers. Solution for all the problems is demonstrated in Figure 3.3.
- **Edge Computing:** Streaming data to a remote storage site and sending the data to the appropriate city authorities is more expensive than using edge computing. In some places, edge computing capabilities such as artificial intelligence (AI) traffic control are already being developed. Intelligent Automation AI is used in traffic management to identify traffic congestion and accidents and provide quick responses to various situations.

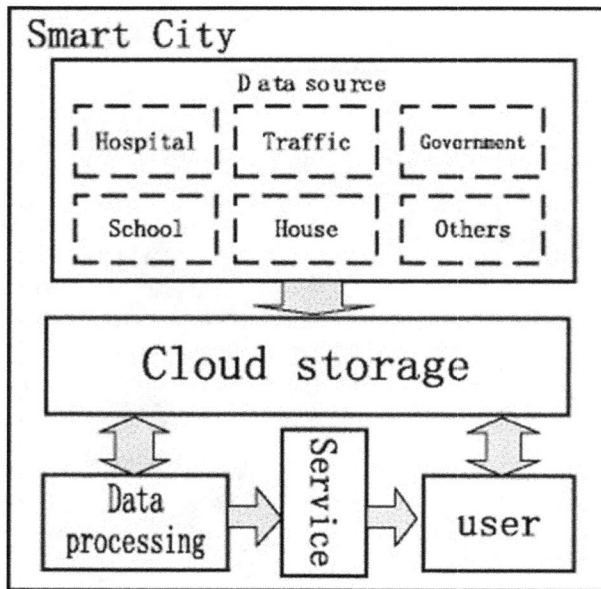

Figure 3.3 Cloud storage in smart cities.

- **Hybrid Storage:** Hybrid data storage systems combine the best of both worlds, allowing for real-time alerts based on conditions and rich data that can generate new results.

3.4 NEED FOR DATA ANALYTICS IN SMART CITIES

Six major reasons for why the data analytics is needed in smart cities is listed below.

3.4.1 Reliability

Security of citizens is a major concern for every society and protecting them under any circumstances is very important. Predictive analytics can be used to look at historical and geographic data to see when and where crime is most likely to occur, in order to stay ahead of the city. When the necessary data makes a city a safer environment, a huge improvement is visible.

3.4.2 Transport

When it comes to big data recovery, the traffic can be handled instantly. Everyone hates traffic, but it is possible to reduce it by using the previous

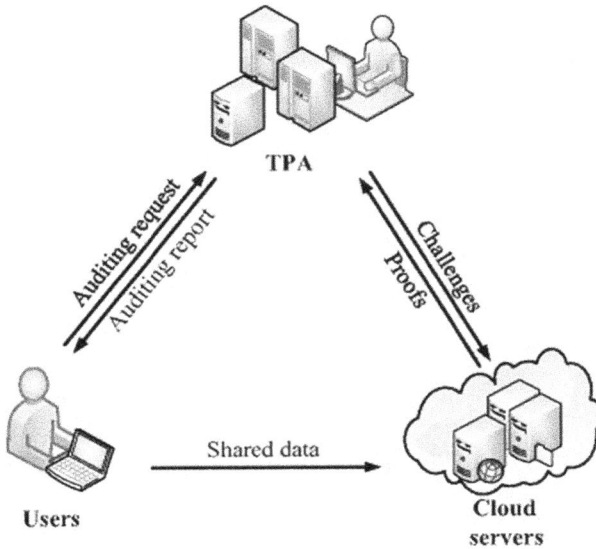

Figure 3.4 Working of cloud.

data properly. You can explore trends in reducing traffic congestion and help traffic authorities develop better ways to manage and monitor traffic within the city by analyzing data received from traffic authorities. Accidents can be reduced through big data analysis.

3.4.3 Planning

Sensors installed in the city provide a clear picture of what is missing in a city and how to improve the current situation. Analyzing the city's current needs using data can help identify areas in need of repair and renovation. Modeling infrastructure needs in a city is easy when using highly accurate data to pinpoint where growth is needed.

3.4.4 Future proofing

Our cities are becoming smarter day by day as a result of increased urbanization and many projects are still underway to fully transform the smart cities of the future. Inside the city, there is technology for intelligent traffic management and real-time administration and monitoring. Data collected from various sources can be used to create a more sustainable environment with improved energy efficiency and reduced resource waste. It is possible to explore the expansion of the current infrastructure and prepare for the future demands of the city using predictive analytics.

3.4.5 Effective spending

A significant amount of money is needed to make the necessary improvements and achieve the desired results of transforming a city. Money is often spent on rehabilitating or redeveloping a city, which is classified as normal infrastructure upgrades. The data obtained in a smart city can identify the most vulnerable places and what kind of renovations are needed using big data analytics. Investments can be made in appropriate domains based on thorough analysis.

3.4.6 Sustainability

Continuous monitoring of a city's growth will provide ongoing development updates, allowing for appropriate improvements to be made as needed. The number of products developed when the technology is implemented will provide a clear picture of the required development, which is one of the keys to long-term sustainability. Data is an important aspect in determining the outcomes of urban development. Changes can be made in the city, but maintaining the current growth is difficult.

3.5 DATA ANALYTICS TOOLS

Data analytics tools are software and programs that collect and analyze data about a business, its customers, and its competition to help improve processes and uncover insights for making data-driven decisions. Some of the data analytics tools are listed below:

- R and Python
- Tableau
- Apache Spark
- QlikView
- Splunk
- KNIME
- Microsoft Excel
- RapidMiner
- Power BI
- Talend

R and Python are popular programming languages used in the field of data analytics. R is an open-source tool used for statistics and analysis, whereas Python is a high-level, descriptive language with simple syntax and dynamic semantics.

Tableau is a market-leading business intelligence tool used to analyze and visualize data in an easy-to-use format. A leader in the Gartner Magic Quadrant 2020 for the eighth consecutive year, Tableau lets you work with live data sets and spend more time analyzing data than data wrangling.

Apache Spark is one of the most successful projects of the Apache Software Foundation, a cluster computing framework that is open source and used for real-time processing. Being the most active Apache project at the moment, it comes with a fantastic open-source community and interface for programming. This interface ensures fault tolerance and implicit data parallelism.

QlikView is a self-service business intelligence, data visualization, and data analysis tool. Named a 2020 Gartner Magic Quadrant Leader for Analytics and BI Platforms that aim to accelerate business value through data by providing features such as data integration, data intelligence, and data analytics.

Splunk is a platform used to search, analyze, and visualize machine-generated data collected from applications, websites, etc. Named a Visionary in ABM by Gartner in the 2020 Magic Quadrant, Splunk has developed products in a variety of fields such as IT, Security, DevOps, Analytics.

At KNIME, creating and producing data science using an easy and intuitive environment enables every stakeholder in the data science process to focus on what they do best.

Microsoft Excel is a platform that helps you gain better insight into your data. Being one of the most popular tools for data analysis, Microsoft Excel offers users features such as sharing workbooks, working with the latest version for real-time collaboration, and adding data directly from photos to Excel.

RapidMiner is the next tool on our list. Being named a Visionary in 2020 Gartner Magic Quadrant for Data Science and Machine Learning Platforms, RapidMiner is a platform for data processing, building Machine Learning models, and deployment.

Power BI is a Microsoft product used for business analytics. Named a Leader for the 13th consecutive year in Gartner's 2020 Magic Quadrant, it provides interactive visualizations with self-service business intelligence capabilities, enabling end users to build dashboards, and reports on their own without relying on anyone else.

Talent is one of the most powerful data integration ETL tools available in the market and is built on the Eclipse graphical development environment. Named a Leader in Gartner's Magic Quadrant for Data Integration Tools and Data Quality Tools 2019, this tool lets you easily manage all steps in the ETL process and aims to provide consistent, accessible, and clean data for everyone.

3.6 DATA ANALYTICS TOOLS FOR SMART CITIES AND TOWNS

Hardware and software are fundamental and important components for fine-tuning data, and first, there are technologies such as Massively Parallel Processing (MPP) architectures that help speed up its processing. However, it is necessary to resort to managing unstructured or semi-structured data for technologies like Map Reduce or Hadoop, which are responsible for managing structured, unstructured, or semi-structured information. Tools used for big data in smart cities should be capable of processing large data sets, massive data, reasonable computation time, and sufficient accuracy. The following tools play a major role in data analytics in smart cities and towns and it is explained as follows:

3.6.1 Hadoop

It is a framework that allows distributed processing of large datasets by groups of computers using simple programming models. A smart city refers to the continued use of technology for the benefit of society. As the city evolves over time, components and subsystems such as smart grids, smart water management, smart traffic and transportation systems, smart waste management systems, smart security systems, or e-governance are added. These components ingest and generate multiple structured, semi-structured, or unstructured data that can be processed in blocks, micro-blocks, or in real-time using various algorithms. The ICT infrastructure must be able to handle increased storage and processing requirements. When vertical scaling is no longer a viable solution, Hadoop can provide efficient linear horizontal scaling, storage, processing, and data analysis in a number of ways. It enables architects and developers to choose a stack according to their needs and skill-levels. In this paper, we propose a Hadoop-based architectural stack that can provide the ICT backbone for efficiently managing a smart city. On the one hand, Hadoop, together with Spark and numerous NoSQL databases and associated Apache projects, is a mature ecosystem. This is one of the reasons why it is an attractive option for smart city architecture. On the other hand, it is very dynamic; things can change very quickly, and many new configurations, products, and options continue to emerge as others fade away. We discuss and compare different products and engines based on a process that considers how the products perform and scale, as well as code reuse, innovations, features, and online support and interest to create an optimized, modern architecture community.

3.6.2 Map Reduce

Map Reduce was designed by Google in 2003 and is considered a pioneering platform for processing massive data, while modifying the resources under the same hardware as the resources under the same hardware. Map Reduce divides the processing into two functions:

1. Map Function
2. Reduce Function

Map Function: Ingestion and transformation of input data and where input registers are processed in parallel. The system processes key-value pairs, reads them directly from the distributed file system, and converts these pairs into other intermediates. User-defined functionality: Each node is responsible for reading and modifying pairs of one or more partitions.

Reduce Function: The master node aggregates the pairs by force and distributes the combined results to the reduction processes at each node. A reduce function is applied to the list of values associated with each key to produce the output value.

3.6.3 Apache storm

It is a distributed open-source system that has the advantage of handling data processing in real time as opposed to Hadoop, which is designed for batch processing. Apache Storm allows the creation of real-time distributed processing systems, which can quickly process unlimited data streams (more than one million tuples can be processed per second), which is highly scalable, easy to use, and guarantees low latency (processing very large amounts of data messages with low latency). It also provides a very simple framework for building applications called topology. A storm is based on a topology consisting of a complete network of peaks, bolts, and flows. A peak is a source of currents, and bolts are used to process the currents to produce currents. Storm can be used for many cases like real-time analytics, online machine learning, continuous computation, and distributed RPC, ETL.

3.6.4 Apache spark

It was born as an alternative to solve graph reduce/Hadoop limitations. It can quickly load and query data in memory, is very useful for iterative processes, and provides a simple programming model that supports a wide range of applications, allowing querying of structured and semi-structured data using machine learning and SQL language. Spark provides more functionality than Hadoop/MapReduce. Ease of use stands out among its main advantages as it can be programmed in R, Python, Scala, and even Java. Spark has its own computed cluster management system, so it only uses Hadoop HDFS for storage.

3.6.5 Apache Flink

Flink is a project of the Apache Software Foundation, developed and supported by a community of more than 180 open-source contributors and used in production at many companies. It is considered an open-source flow processing framework that allows real-time transactional analysis of large volumes of data with a single technology.

Flink allows programmers great flexibility to associate events with different concepts of time (event time, injection time, execution time); it offers low latency, high performance, multi-language APIs, irregular events, fault tolerance, and consistency.

3.6.6 Flume

It is a commonly used injection or data collection tool for Hadoop. Flume is a distributed, reliable, and available system that collects and transfers data from various sources to a centralized data warehouse such as Hadoop Distributed File System (HDFS). Some of its functions include fault tolerance, adaptive reliability mechanism, and fault recovery service. Flume relies on a simple extensible data model to handle massive distributed data sources. Although Flume complements Hadoop well, it is an independent component that can work with other platforms. He is known for his ability to run multiple processes on a single computer. Using Flume, users can send data from multiple high-volume sources (such as Avro RBC Source and Syslog) to sinks (such as HDFS and HBase) for real-time analysis. Additionally, Flume provides a query processing engine that transforms each new batch of data before sending it to a specific recipient.

3.7 CONCLUSION

Big data plays an important role in smart city's processing of data collected by IoT devices so that further analysis can be done to identify the trends and needs of the city. Sensors deployed across the city generate huge amounts of data, but if used efficiently, it can lead to many changes. Cities can identify trends and needs by analyzing data from IoT devices and sensors. Analytics can help drivers find parking space and reduce road accidents and congestion. The data could also help reduce crime, smart city lighting, and water and electricity systems.

REFERENCES

1. https://www.researchgate.net/publication/325672561_Big_Data_Tools_for_Smart_Cities
2. https://www.edureka.co/blog/top-10-data-analytics-tools/

3. https://www.investopedia.com/terms/d/data-analytics.asp
4. Singh, P., Arora, K., Rathore, U.C.: Control strategies for improvement of power quality in grid connected variable speed wecs with dfig–An overview. *Journal of Physics: Conference Series*, 2022.
5. VMware: Apache flume and apache sqoop data ingestion to apache hadoop clusters on VMwarevSphere. TechnicalWhitePaper. VMware, 2013.
6. Alejandro, P.G.: Apache Spark – Big Data, pp. 4–6, 2016.
7. Ignacio, D.E.: *Instalación y configuración de herramientas software para BigData*, pp. 17–18. Universidad Politécnicade Valencia, 2016.
8. Friedman, E., Tzoumas, K.: *Introduction to Apache Flink*, 1st–17th edn, O'Reilly Media, 2016.
9. Carbone, P., Ewen, S., Haridi, S., Katsifodimos, A., Markl, V., Tzoumas, K.: ApacheFlink: Stream and batch processing in a single engine, pp. 28–29, 2015.
10. Camargo Vega, J.J., Camargo Ortega, J., Joyanes Aguilar, L.: Knowing the Big Data. *Revista Facultad de Ingeniería*, vol. 24, no. 38, pp. 63–77, 2015.
11. García, S., Ramírez, S.G., Luengo, J., Herrera, F.: *BigData: preproces amiento y calidad de datos*, pp. 18–20. University of Granada, 2016.
12. Rodriguez Sánchez, F.M.: *Herramientas para Big Data: Entorno Hadoop*, pp. 9–13. Universidad Politécnica de Cartagena, 2014.
13. Guerrero, F.A., Rodriguez, J.E.: *Diseño y desarrollo de unaguía para la implementación deun ambiente Big Data en la Universidad Católica de Colombia*, pp. 32–40. Universidad Católica de Colombia, 2013.
14. Arora, K., Kumar, A., Kamboj, V.K., Prashar, D., Shrestha, B., Joshi, G.P.: Impact of renewable energy sources into multi area multi-source load frequency control of interrelated power system. *Mathematics*, vol. 9, p. 186, 2021.
15. Oussous, A., Benjelloun, F.Z., AitLahcen, A., Belfkih, S. Big Data technologies: A survey. *Journal of King Saud University – Computer and Information Sciences*, vol. 30, pp. 431–448, 2017.
16. Arora, K., Kumar, A., Kamboj, V.K., Prashar, D., Jha, S., Shrestha, B., Joshi, G.P.: Optimization methodologies and testing on standard benchmark functions of load frequency control for interconnected multi area power system in smart grids. *Mathematics*, vol. 8, p. 980, 2020.

Chapter 4

Industrial Internet of Things and its applications in Industry 4.0 through sensor integration for a process parameter monitor and control

R. Kiruba, K. Srinivasan, V. Rukkumani and V. Radhika
Sri Ramakrishna Engineering College, Coimbatore, India

CONTENTS

DOI: 10.1201/9781003407409-4

4.1 INTRODUCTION

Measuring and controlling of a parameter plays a vital role in each and every industry. There are numerous industries all over the world, where different parameters in those industries are to be measured and these parameters are to be placed within their limit set by the operator. Among the various parameters, level measurement is very important in industries where level is to be maintained in its limit for the safe operation. So, it is necessary to monitor and control the level in many processes. Nowadays, everything in the world is getting digitalized and is monitored anywhere in the world, that is, even in remote places. This proposed work will help to automate and monitor any level process from any part of the world. This is done by cloud computing. Connecting devices with the internet enables monitoring of any parameter as an easy task.

In general, industries are using PLC to automate a process in industries, but there are many difficulties in automating it. In PLC, there is still a lot of wire to cooperate with it. There is a need for an observer and analyst to monitor the process parameter involved in it. But in some chemical industries, humans cannot go near the process since it can be harmful; in such cases, it is not applicable. The main aim of process control automation was to monitor the process value and control the values without human interface and replace them with programmed electronic systems.

In existing industries, the level process station was monitored remotely and controlled using an HMI-SCADA environment. The control of the SCADA process has a great demerit of human requirement near the level process to monitor the process. Another system was used in which Raspberry Pi, instead of SCADA system, is used to control the level process. A system was used for monitoring and controlling of level process from remote places using Raspberry Pi. Raspberry Pi was opted in this proposed system as the main node for measuring and controlling a level process. In this method, the data from different sensing devices has been integrated, which can communicate through a single network on which the proposed system was connected; by utilizing this method, it doesn't need external controllers for taking the decision to control the actuators where cloud will itself act as a controller. This method forms the backbone of the proposed methodology.

IoT is an emerging technology which can connect many things to the internet. Connecting devices with internet enables these devices to collect data and send them to cloud. IoT has a greater advantage in future, which can automate the whole world. Using the protocol, we can communicate with other devices. It seems that numerous devices are integrated to the internet. Automation is the main future work of IoT.

4.2 PROBLEM STATEMENT

Initially, process stations were using Human Machine Interface system like SCADA for monitoring the system and controlling the process variable to the set point. In order to overcome the difficulties in the SCADA system, the Raspberry Pi has been introduced for controlling in recent days. The existing systems in the process industries using Raspberry pi were effective to monitor and control the level process in remote areas. This allows devices to be aware about web so that monitoring and controlling was done by a web browser. This method log processes data and monitors and controls it via mobile application. But the system needs an external controller like Raspberry Pi to control the process. In order to improve the method, a new system was proposed, in which it does not require an external controller to control the level process. Instead, cloud itself acts as a controller.

4.3 OBJECTIVE

The major objective of this project was to monitor and control the level process through a cloud controller in order to avoid external controllers. The variations in the level process parameter will be sent to the monitoring device so that the changes in the process variables can be monitored continuously. The proposed system helps in maintaining level within an optimum range when it is needed. The data from the level process station was sent to the cloud server via ESP8266. Set point is given to the cloud server by using a monitoring device, say a mobile phone. Programming was done in cloud, say amazon, using eclipse java platform. Data are fed into the cloud server with greater speed to increase distance from remote areas and received at the desired point.

   The data from the level process are then processed as per the program in cloud and controlled. Hence, cloud acts as a controller and sends the control action back to ESP8266. Thus, the pump is turned off through the control signal from cloud. Broadband internet is widely accessible due to the expansion and efficiency of the telecom industry. With the development of technology, it is now less expensive to make necessary sensors, and since external controllers are no longer required, networking devices are also less expensive.

4.4 RELATED WORKS

Although the World Wide Web had its origins in the 1960s, companies were not able to use it until the mid-1990s. The World Wide Internet was created in 1991, and the Mosaic web browser, which allowed users to view web pages with both text and images, was introduced in 1993. An entirely new business category known as an Application Services Provider, or ASP, started to emerge as Internet connections became faster and more dependable. The consumer would pay a monthly charge to access the program through the Internet, while the ASP would purchase the computer hardware and maintain it. It is known as cloud computing. The IoT can link internet-connected gadgets that are integrated in diverse systems. When gadgets or objects can digitally represent themselves, they are controllable from any location. Following that, the connectivity enables us to collect more data from more sources, providing additional opportunities to boost productivity and enhance safety and IoT security. This section examines the IoT research that has already been done and the different controller approaches that have been used for processing control strategies in various sectors. There will be several metrics for crucial operations that need to be watched in a production setting. A process control system keeps an eye on the setting in the factory. A central control system serving as a server will receive real-time values of these parameters. Through feedback systems, these data are compared to the predetermined set-points, and any required warnings are generated on the display system so that remedial action may be performed.

"Integration of Cloud Computing (CC) with Internet of Things: A Survey," by Botta et al., is a review of the research in this area. Assessing the fundamentals of both IoT and CC first, we then explored how they complement one another and what is now propelling their integration.

Gang Zhao (2011), "Wireless Sensor Network for Industrial Process Monitoring and Control", explained about the smart industrial application using IoT, and complete automation is done to monitor the process by using ZigBee module. This information is updated to cloud through one or more channels using ESP8266 Wi-Fi module [34].

In "Personalized Health Monitoring System utilising IoT and Cloud," S. Ramamoorthy (2017) detailed on how patient health is constantly tracked and the collected data is sent to wired or wireless sensor networks. The patient's input may be analyzed by health sensors, and the findings of all the parameters are saved in a cloud database. Both patient's and the doctor's information is kept in the cloud database [95].

The data recording capabilities of LabView are highlighted in the article by Helia Guerra et al. from 2021, "Demonstration of Programming in Python Using a Remote Lab with Raspberry Pi," which also uses a graphical user interface to evaluate all sensors and valves with the system. Then, a brief description of the various techniques used to integrate level process measurement systems is given, along with a list of the many model controller parameters created for level monitoring.

In their study, "Wireless Sensor Node to Monitor Industrial Environment Parameters," Mane Deshmukh et al. (2020) provided examples of how to build and create a wireless sensor node for a wireless sensor network in order to monitor industrial parameters. A cutting-edge technique for monitoring a large region is the wireless sensor network (WSN). The WSN, which uses the MQ-3 to detect alcohol gas, is created with a focus on monitoring the ambient alcohol gas concentration.

Nerella Ome et al. (2016), "Internet of Things based Sensor to Cloud Computing System using ESP8266 and Arduino Due," carried out an analysis on the requirements for wireless in process automation and the implementation of WSN technology on industrial process monitoring and control with a review of the advantages of adopting WSN technology for industrial control.

4.4.1 Inferences through the survey

Process control approach and its many strategies were researched. A presentation of online experiences using a Raspberry Pi that is connected to a wireless sensor network and allows users to engage with the physical environment through real-life instances was given. Different CC methodologies were investigated.

4.5 EXISTING METHODOLOGY

This chapter deals with the detailed explanation of block diagram and the components used in the existing methodology. In this method, the level process station was interfaced with SCADA system, which does not have the feature of remote access and data storage in server. The block diagram explanation of the existing system is discussed.

4.5.1 Level process station

Level process station controlled using SCADA as HMI as shown in Figure 4.1 has been explained for better analysis of basic level control.

The level setup comprises a main supply water tank from which water is fed into the tank in the process station as the input with the aid of a pump. The amount of inlet flow of water is measured by a rotameter. The level transmitter used for level sensing is fitted on a transparent process tank. The process parameter, which varies with respect to the input to be controlled by a SCADA controller, produces the current signal, which could be fed into an I/P converter to convert to the pneumatic signal, which in accordance controls the pneumatic control valve acting as an actuator. The controller is connected to HMI through USB port for monitoring the process.

4.5.2 Monitoring and control of level process

There are numerous methods to monitor and control the level process. Popular methods which are widely used in most of the process industries are listed below.

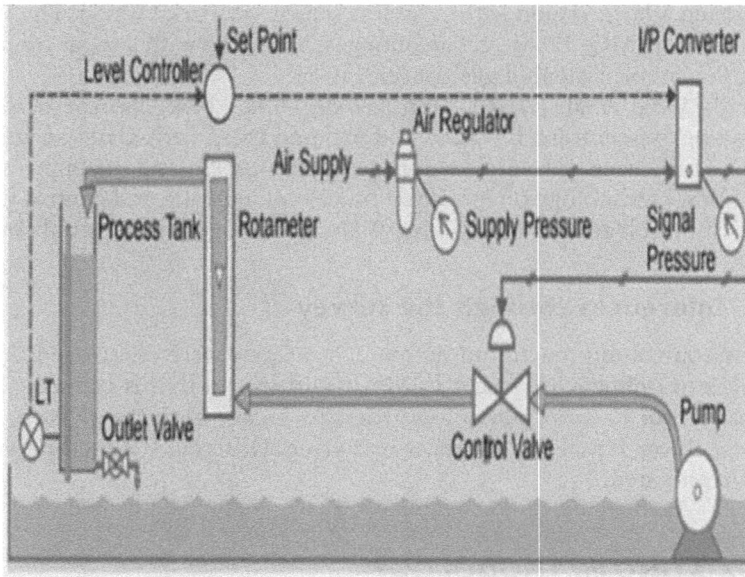

Figure 4.1 Level process station using SCADA.

4.5.3 SCADA

In several industries, supervisory control and data acquisition (SCADA) was utilized to monitor and regulate process levels. Any process may be controlled and data acquired with SCADA systems. HMI are sometimes referred to as local user interfaces because they let process engineers control the process on-site and perform SCADA programming to tailor the system. However, this technique of control has the drawback of requiring an observer to keep an eye on the process.

4.5.4 Level process monitoring and control through SCADA replacement

Improvements were made in the above existing method to meet the specified requirements in today's digital scenario, which can even monitor and control any process station parameter even in a remote place from anywhere in the world by cooperating the idea of IoT as shown in Figure 4.2. In this improved methodology, the station uses Raspberry Pi as a controller instead of the SCADA HMI system. Raspberry Pi acts as a central processing unit of the entire system. This is much adoptable for many industries which handle multiple parameter controlling or monitoring systems. This is a miniature of the actual CPU (Central Processing Unit) used in PCs. This has inbuilt USB (Universal Serial Bus) and Ethernet ports and also consists of HDMI

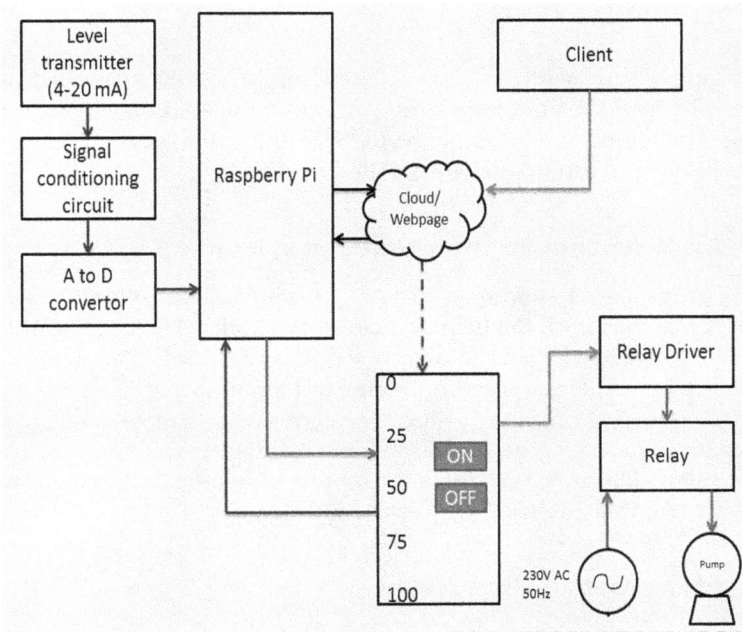

Figure 4.2 Block diagram of existing methodology.

(High-Definition Media Interface) port through which it can be connected to the display units. It is to this Raspberry Pi unit that the level transmitter value from the ADC is sent. This sent value to the Raspberry Pi is then processed with the preloaded algorithms. These algorithms are mapped in such a way that the process value and the set point evenly match the algorithm value, which goes on, and when it does not match the processor, it chooses an alternative to bring the process value to the set point value.

The pump feds water to the inlet pipe from the tank present in the level process. This water then passes through the rotameter where flow of water can be identified. The water from the rotameter was finally fed to a transparent tank. The level of the tank was measured by a level transmitter. The output of the level transmitter was measured in milli amperes (mA), but Raspberry Pi accepts only voltage input. So, an I to V converter is used in between Raspberry Pi and the level transmitter where the input current is converted into voltage. The analog input is converted into digital output using ADC. SPI communication has been used to Interface the Raspberry Pi with MCP 3008.

Python programming is done in the Raspberry Pi processor according to the desired set points. The fetched data are sent to the Cloud server that store the data. Thus, the user can monitor the level parameter from anywhere using software in mobile applications.

4.6 PROPOSED SYSTEM

This chapter deals with the detailed description of the proposed system design. This gives a brief explanation of components used and the entire process development. This explains in brief about the working of the proposed system and different elements used in it.

4.6.1 Introduction to the proposed system

In the proposed methodology, the level monitoring system shown in Figure 4.3 is done with the help of a cloud controller. The level is measured by a level sensor and then the data is fed into the cloud. The data is given to the cloud via ESP 8266(wi-fi module) and monitored using a monitoring device. In this method, cloud is used for acting as a controller to store and react to set point data automatically. Table 4.1 depicts the components used in the project. This method admits the user to access the data at anytime and anywhere from the internet.

4.6.2 Flowchart of the system

The mechanism of the proposed system is shown in Figure 4.5.

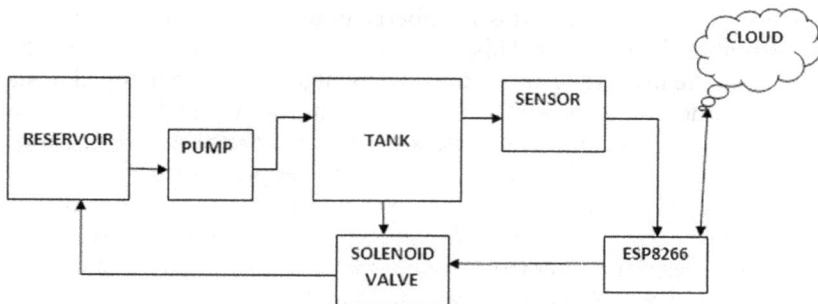

Figure 4.3 Block diagram of IoT-based process controller system.

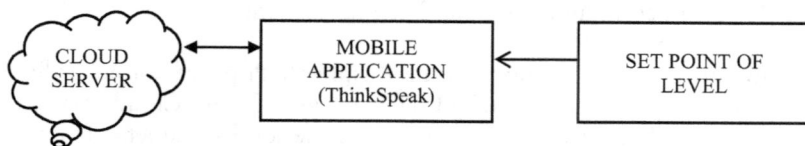

Figure 4.4 Monitoring and control of level parameter through cloud.

Table 4.1 List of products used in the proposed system

Product	Specification	Function
ESP8266	Wi- Fi Module, Supply Voltage 3.3V	Receives the control signal from cloud and turns ON/OFF the pump
Pump	12 VDC, Capacity: 300 l/h	To pump the water from reservoir to the tank
Float Type Level Sensor	LF 250	Sensing the level of the tank
Solenoid valve	Normally Close, 24V DC	To take back the excess water from the tank back to the reservoir.

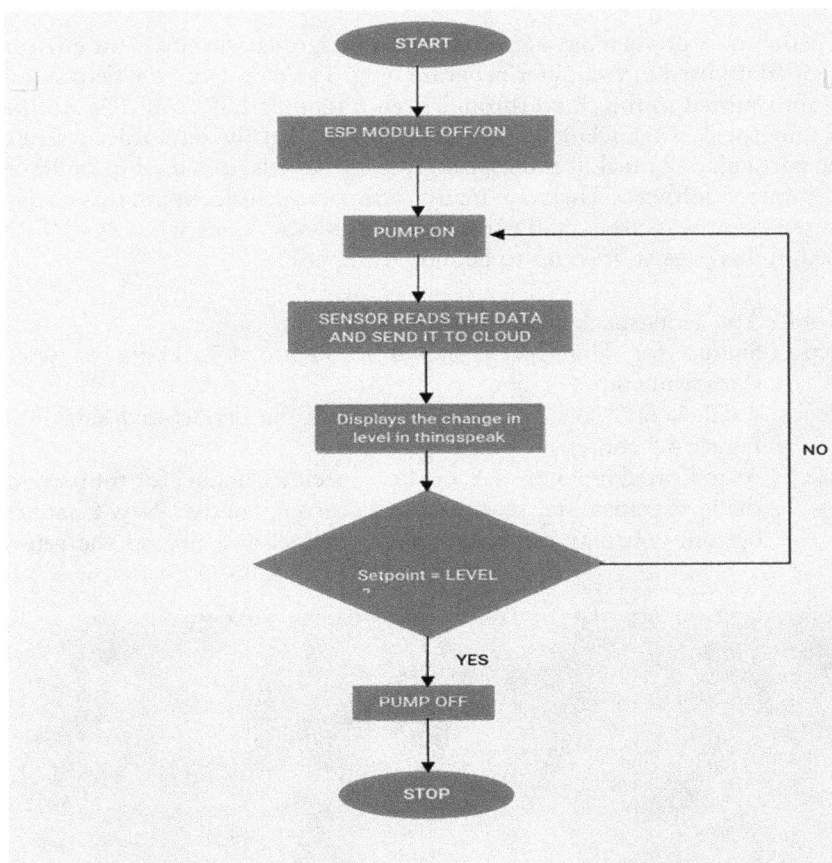

Figure 4.5 Flowchart of the proposed.

Figure 4.6 Eclipse software.

4.6.3 Eclipse

It is the most popular Java IDE and is an integrated development environment (IDE) used in computer programming. The data from the field should be transmitted to the cloud through a wi-fi module ESP8266. The API key of thingspeak was included in the program, so that the data can be sent to the particular channel in thingspeak. In other words, it is used to build our customized software. There are many versions in eclipse, but in this project, eclipse neon was used as illustrated in Figure 4.6. Since windows 10 was used in this project, it seems to be more compatible.

Mobile App Thingspeak creation steps:

Step 1: Signup for Thingspeak shown in Figure 4.7, Login to www.thinspeak.com

Click on "Sign Up" icon to furnish the details and sign in as Figure 4.8 shows

Step 2: As indicated in Figure 4.8, create a specific Channel for transferring the level transmitter sensor data. By turning on the "New Channel" option, you may establish a new channel and upload the sensor

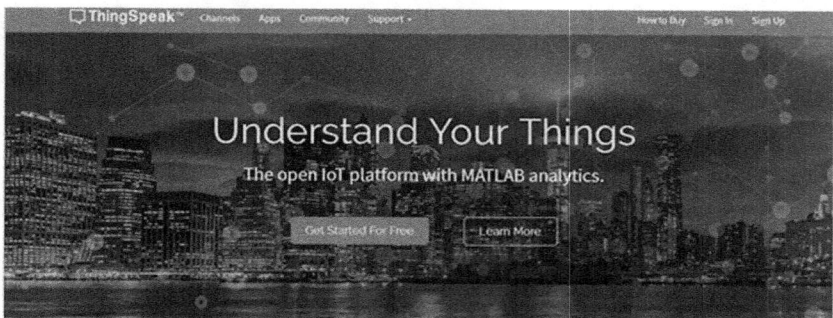

Figure 4.7 Thingspeak platform.

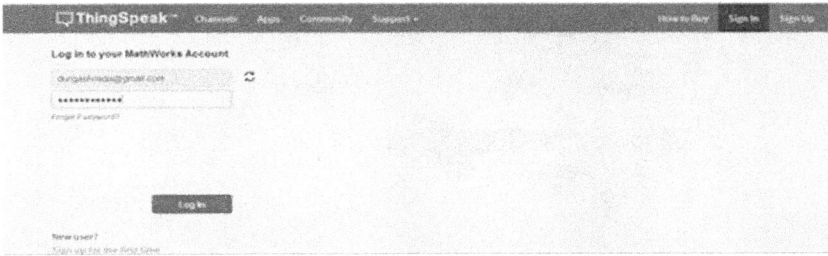

Figure 4.8 Signing in to the Think speak website.

readings to the server. Enter the Description of the material that will be uploaded when the "New Channel" page has loaded. You can provide the name of the data, such as level, in field 1. If more Fields are necessary, tick the box next to the Field option and type the relevant data name.

In order to save of all settings, check on "Save Channel" button.

Step 3: Get an API Key. To upload data, an API key is need, which will later be included in a cloud program to upload sensor data to Thingspeak.

Once the "Write API Key" is enabled. It is almost ready for an app to upload data, except for the python code.

Step 4: Check Thingspeak API and Confirm data transfer Open your channel and you should see the level uploading into thingspeak website.

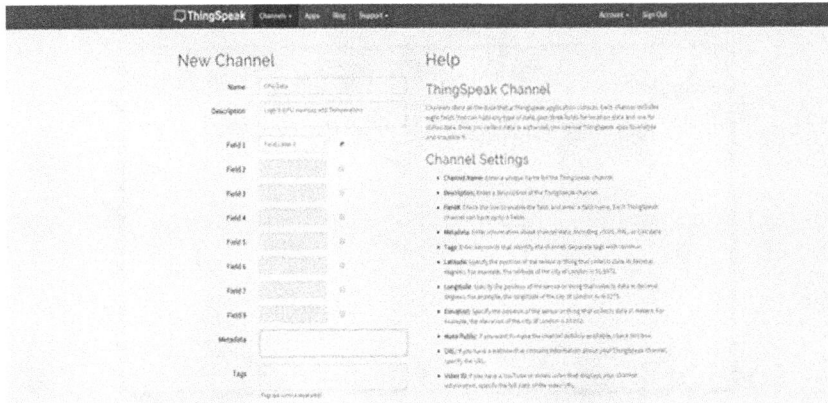

Figure 4.9 New channel creation.

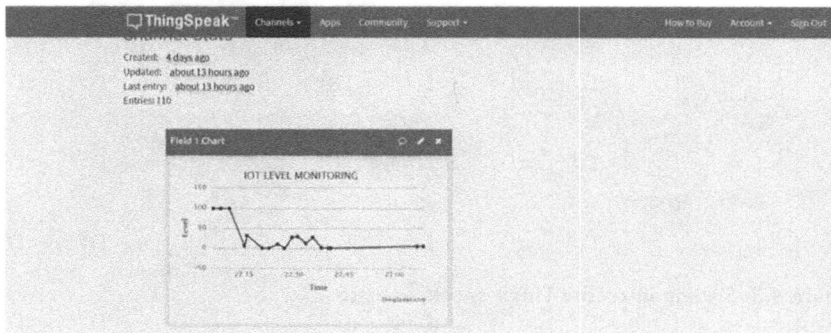

Figure 4.10 Creating field charts to display data.

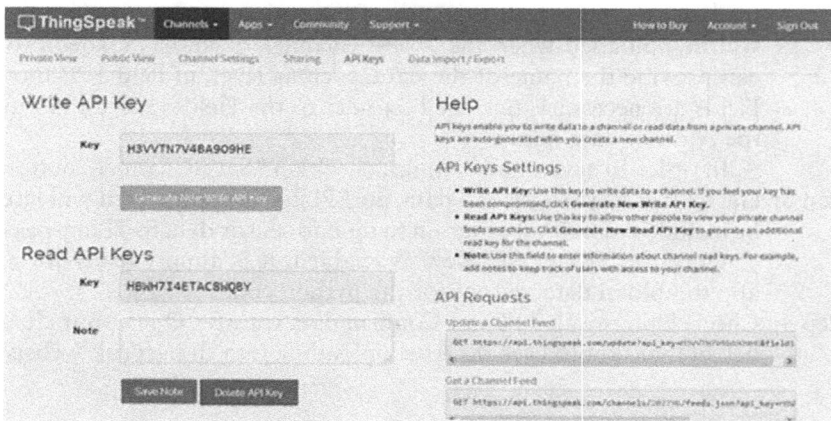

Figure 4.11 To copy the write API key of the channel.

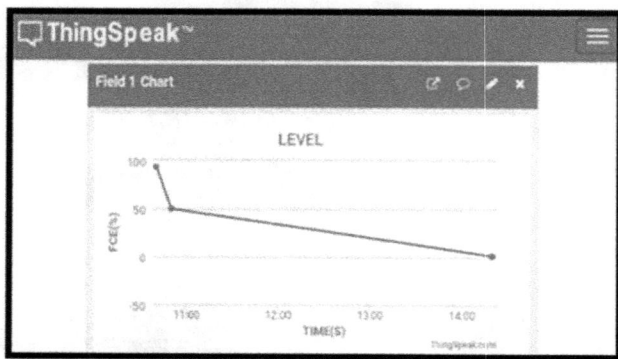

Figure 4.12 Data displayed in field chart.

4.7 CLOUD DESCRIPTION

Cloud computing is an emerging technique, which utilizes Internet to create a network of different servers. This replaces the use of hardware and provides a virtual memory, which stores data, manages data, and reacts with high speed and low cost.

In this project, the input parameter corresponds to level changes with respect to time. So, the static cloud which has a standard data storage size cannot be used, as it cannot adopt to variations in input parameters has to numerous data. Obviously, Dynamic cloud was used here to record and retrieve the level from the real process that varies in time.

4.8 TRAIL RESULTS

This section explains about the analysis of a level process parameter and also about the performance characteristics of the process station in industries. Table 4.2 gives the brief experimental results, and the characteristics curve is shown with respect to the set point and control action. The data are then sent through Internet to the thingspeak mobile App where the system can be accessed even in remote places acting as Industrial Internet of Things.

The set point is inputted through a mobile App Think Speak and the data that is driven from the real-time working process from level tank is fed to the cloud acting also as a controller via ESP8266 Wifi Module. When the data or parameter of level, which has been measured from sensor of a real-time process exceeds the point level set by the operator through the mobile application, then, the appropriate control action according to the error will

Table 4.2 Output analysis

TRAIL	Set point	Control action
1	16	TAKEN
2	21	NOT TAKEN
3	17	TAKEN
4	29	TAKEN
5	47	TAKEN
6	97	TAKEN
7	72	TAKEN
8	84	TAKEN
9	74	TAKEN
10	76	TAKEN
11	24	TAKEN
12	86	TAKEN

send a control signal, which is passed from cloud to the pump to turn off the pump. Table 4.2 shows the number of trails carried out and a few of them are explained below based on the set point and control action taken or not.

4.8.1 TRAIL I

Here, the set point was dictated as 16 through a mobile device application Think speak and the data are collected from the Level Transmitter, which was

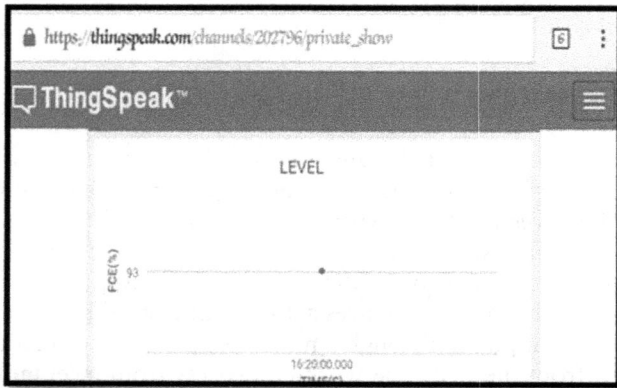

Figure 4.13 Trend of level process variable.

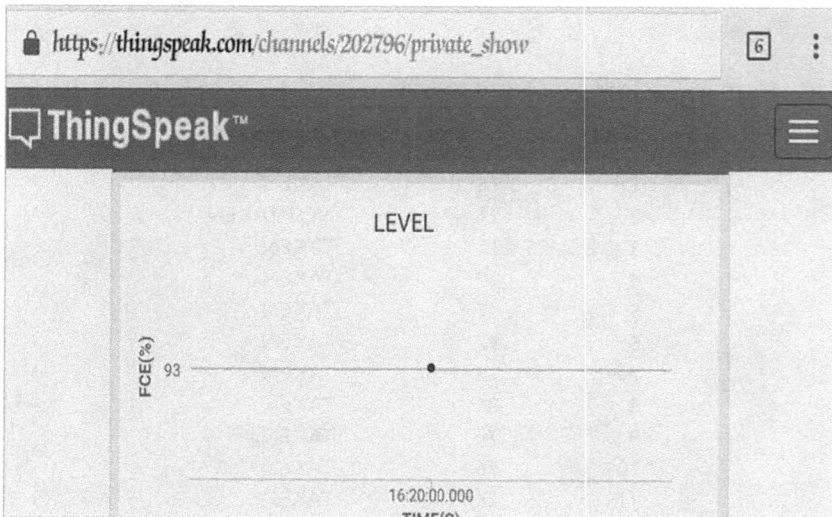

Figure 4.14 Trend chart of process variable (trail 3).

updated in the cloud server, and once it reaches the set point, the pump will be turned off via the control signal from a decision taken by the cloud controller.

4.8.2 TRAIL 2

The set point 21 mm of level was set as a targeted value via the mobile device, and simultaneously, the data from level transmitter was updated. When the data read from the sensor value tracks the set point, then the control signal reaches the pump playing the role of an actuator to turn it off. But the pump is not turned off when the set point is reached. There is an offset value of level in the controller.

4.8.3 TRAIL 3

The level set point 17 mm was given via the mobile device, and simultaneously, the data from the sensor was loaded in the cloud. When the read value from the sensor value attains the set point, then the control signal reaches the pump to turn it off as if the desired level has been attained.

4.8.4 TRAIL 4

Once we turned ON the setup, a set point was given as usual. But the sensor reading is not uploaded in thingspeak website.

While troubleshooting, we analyzed that the sensor got damaged and it stopped working. It was replaced with another one. And the remaining trails are taken. Performance of the characteristics of the system is shown in Figure 4.15.

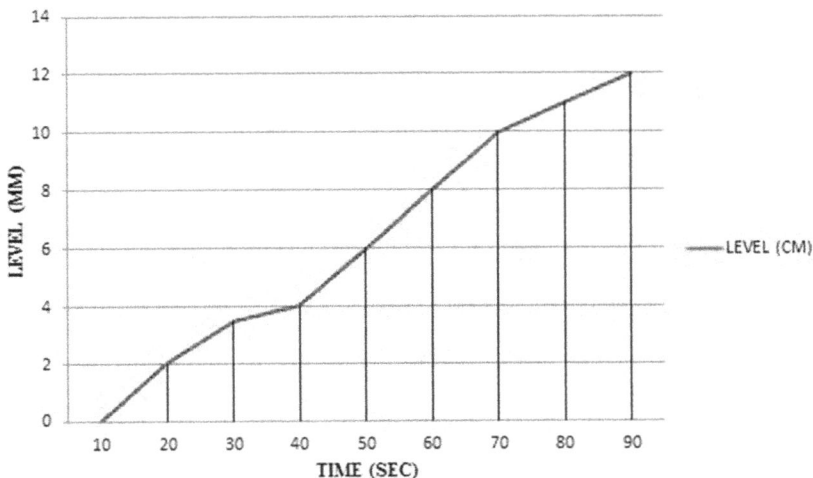

Figure 4.15 Characteristic performance of level process through Mobile App.

4.9 PROJECT SETUP

The prototype of level process, which works with the IoT, plays a role of miniature Industrial IoT. There is a source tank; from this tank, the process station tank will be fed, and its level has been measured by the level transmitter. This transmitter in turn in terms of current value will be sent to the current to pressure converter. This pressure value is set according to the level set point; the control action in the cloud will be taken according to the level value sensed and the point set by the user. This proposed system is shown in Figure 4.16.

4.10 INFERENCE

The overall performance of the process is good, and in general, there is a 10-second delay. But at the same time, the control action taken was too delayed due to the slow response of Amazon cloud server. Generally, the response of cloud depends upon network speed. Since in our area where the project is operated, the network connectivity was slow at times, so the response was tedious sometimes.

4.11 CONCLUSION

The level is one among the various important parameters measured in industries. Here, in this proposed system, level process parameter is the crucial

Figure 4.16 Proposed system.

physical variable of the industrial process automation and IoT is taken into account for monitor and control. Implantation of this project using only Wi-Fi module ESP8266 for not only level monitoring or any other physical variable can also be used as a method to monitor process industrial plant, which is designed here for real-time application. It mainly supports online supervision of level process not only within private network (LAN) but also in public network, and can be extended to the wide area network (WAN). The main reflection drawn from this study is that a cost-effective level monitoring and controlling system within cloud using eclipse platform was built, as there is no use of a stand-up controller. This methodology has improved its shortfalls much better with remote access in monitoring of level with this kind of advanced cloud controller.

4.12 LIMITATIONS OF THE PROPOSED METHODOLOGY

1. As discussed in result analysis, there is delay in control actions.
2. The privacy is the major problem in cloud computing. But still, it can be overcome with some advanced future works like cyber security encryption.
3. Since the data are more, it requires a larger server in future.

ANNEXURE-I

```
package embed;
importjava.util.HashMap;
importjava.util.Map;
public class Embed {
static String method = "GET";
public static void main(String[] args) {
//    Http_Reqobj_sndReq = new Http_Req();
      Map<String, String>params = new HashMap<String,
          String>();
      String apiURL = "http://api.thingspeak.com/update?";
      String apikey = "49HW6GLURZD4FWPS", field1 = "020";
String[] response;
try
{
response = Http_Req.sendHttpRequest(apiURL + "api_key=" +
   apikey + "&field1" + field1, method, params);
if (response != null &&response.length> 0)
 {
for (String alMOB1 : response)
{
//                      respUp += alMOB1;
```

```
System.out.println("RESPONSE FROM: " + alMOB1);
            }
        }
    } catch (Exception ec)
{
System.err.println("Ex:" + ec.toString());
    }
        // TODO code application logic here
    }}
```

REFERENCES

Arora, K., "Internet of Things-Based Modernization of Smart Electrical Grid", *Smart Electrical Grid System*, CRC Press, pp. 1–14, 2022.

Pasha, S. "Thingspeak Based Sensing and Monitoring System for IoT with Matlab Analysis", *International Journal of New Technology and Research (IJNTR)*, Volume-2, Issue-6, June 2016, 64–71.

Ramamoorthy, S., "Personalized Health Monitoring System Using IOT and Cloud", *International Journal of Computer Science Trends and Technology (IJCST)*, Volume-5, Issue-3, May–Jun 2017, 156–163.

SafiriyuEludiora, "A User Identity Management Protocol for Cloud Computing Paradigm" *International Journalof Communications, Network and System Sciences*, Volume-4, 2011, 152–163. doi: 10.4236/ijcns.2011.43019. Published Online March 2011 (http://www.SciRP.org/journal/ijcns).

Chapter 5

Automated computer-aided diagnosis of COVID-19 and pneumonia based on chest X-ray images using deep learning

Classification and segmentation

Maria C. Moreno, Brayan Daniel Sarmiento, Daniel Steven Moran, Fabián Enrique Casares and Oscar J. Suarez

University of Pamplona, Colombia

CONTENTS

5.1 INTRODUCTION

The novel coronavirus disease (COVID-19), caused by SARS-CoV-2, represents a large family of viruses that cause diseases associated with the respiratory system. After entering through the respiratory tract, it critically affects the patient's lungs, developing infectious symptoms that range from cold to severely affecting the lungs causing pneumonia, difficulty in breathing, acute renal failure, multiple organ failure, and even death.[1] Likewise,

DOI: 10.1201/9781003407409-5

51

susceptibility to the virus has been demonstrated by numerous animal species. Therefore, it is estimated that in different cases, the virus was able to evolve and infect the human species, causing rapid and easy spread among the population.[2] The first COVID-19 case was reported in Wuhan, China, in December 2019, and on January 30, 2020, the Coronavirus outbreak was declared a Public Health Emergency of International Concern by the World Health Organization (WHO).[3] Finally, on March 11, 2020, the WHO characterized the situation as a pandemic due to the rapid spread of the virus in multiple countries around the world. A detailed analysis confirms that the increase in the growth rate of registered cases has generated the collapse of the health system in various developed countries, some of the most affected was the United States with 560,433 active cases and 22,115 deaths, Spain with 166,831 active cases and 17,209 deaths, and Italy with 156,363 active cases and 19,899 deaths.[4] Moreover, the statistics show that the number of deaths recorded in Italy on March 19, 2020, surpassed the number of deaths recorded in China.[5] Accordingly, the presence of the virus has become a trigger in the increase in the mortality rate in the world, whose current number of deaths is around 244,000 registered cases.[6]

On the other hand, some considerations on asymptomatic infection by COVID-19 and its implication in the spread of the disease have been exposed, becoming one of the main sources of debate and concern for the scientific community. This led the international community to search for new ways to reduce the spread of the virus and avoid situations of collapse in the health system.[7] As a consequence, many countries adopted social distancing as one of the biosafety measures to reduce contagion peaks.

Currently, multiple tests have been standardized for the diagnosis of virus infection in patients through sputum or blood sample analysis; reverse transcription polymerase chain reaction (RT-PCR) is one of the most reliable and efficient methods. The main purpose of RT-PCR is the detection of viral nucleic acid in sputum or nasopharyngeal swab. However, this mechanism is an expensive method involving long periods of time for its validation in laboratories, whose nature is cumbersome and uncomfortable for patients. In addition, manipulation in taking samples for this type of test requires a level of biosafety with materials and supplies are not easily acquired.[8,9]

Several approaches on COVID-19 suggest the need to generate new alternative diagnostic methods that allow addressing the current problem and stimulating a sustainable ecosystem based on the application of artificial intelligence in which contact approaches are minimized.[10] The use of convolutional neural network (CNN) enables a drastic reduction in testing time using radiology images.[9] Additionally, the deep learning application allows for training the weights of the neural networks on large datasets as well as fitting the weights of the networks with previous training on small datasets.[11]

Starting from these considerations, this study proposes an efficient, low-cost and versatile deep learning based approach for automated computer-assisted diagnosis of COVID-19 and pneumonia using segmentation and classification techniques on chest X-ray images (due to the high degree of similarity between pneumonia caused by COVID-19 and traditional pneumonia). At first, a bank of chest X-ray images is generated to later categorize into a group of covid, pneumonia and normal images. Subsequently, the structure of the neural network is developed which enables diagnosis with minimum requirements. Finally, a portable version for mobile devices based on Flutter and TensorFlow libraries is made.

5.2 LITERATURE REVIEW

COVID-19 using artificial intelligence has been the focus of multiple researchers and institutions to alleviate the work pressure of front-line radiologists and contribute to the control of the epidemic through early diagnosis of the virus, isolation preventive and effective treatment of disease in patients.[12–15] In this section, three fields are addressed for the application of artificial intelligence for the diagnosis of pulmonary pathologies.

5.2.1 Medical image segmentation

Medical imaging plays a major role in disease diagnosis in multiple medical fields, ranging from lung cancer screening to prostate cancer staging. Clinical care has broadened its interest in the application of Machine Learning techniques for the extraction of features that allow providing critical information about the internal state of the patient through computed tomography (CT), magnetic resonance imaging (MRI), digital mammography, among many others imaging in remote modalities, without the need to perform invasive processes.[16] Allowing to automate or semi-automate the delimitation of structures and regions of interest is the anatomy of a patient.[17]

5.2.2 Computer-aided detection and diagnosis

Computer-aided detection and diagnosis (CAD) of pulmonary pathologies based on X-ray images is a field of computer science research that was developed in the 1960s and has been under development in recent decades. It has become one of the most important topics in medical imaging research.[18] In line with this, some contributions related to disciplines that will allow rapid automatic detection of COVID-19, cancer and other lung diseases from radiological images, such as chest radiography and computed tomography, are presented. For example, in a study,[19] an automated deep transfer learning-based approach for the detection of COVID-19 infection in chest X-rays with an extreme version of the inception model is proposed. In another example,[20] the application of transfer learning to a CNN known as AlexNet for the recognition of COVID-19 in chest X-ray images extracting significant biomarkers related to the disease is described. At last, in another study,[21] a deep learning-based method for classifying cancer data from microarrays organized into a set of classes for further diagnostic purposes is exposed.

5.2.3 Annotation-efficient

Recently, different studies have presented on implementing the segmentation and classification of medical images through deep learning techniques for the rapid detection of COVID-19, pneumonia and other diseases in patients. In a study,[22] a deep CNN system for automatic diagnosis of odontogenic cysts and tumors through landmarks in cephalometric X-ray images is presented. Also, in another study,[23] a neural model is proposed for the automatic detection of pneumonia using deep transfer learning through a U-NET type architecture and two classes for classification: Normal and Abnormal. In contrast, in another study,[24] a defect detection approach in welding works by image segmentation based on a Chan-Vese model is described. In addition, in another study,[25] the use of Stenosis-DetNet for the segmentation and identification of stenosis in coronary arteries is proposed. Finally, in another study,[6] a CNN model for the rapid detection of COVID-19 lung infection in chest X-ray images is presented.

5.3 MATERIALS AND METHODS

The proposed methodology followed in this research consists of four main phases: (1) acquisition of the chest X-ray images for the creation of the dataset; (2) application of segmentation techniques for image pre-processing; (3) CNN training for image classification; and (4) development of a graphical user interface for easy interaction during the diagnosis. In this section, a brief presentation of the entire dataset used for model training is made, and all the processing performed on the images to obtain better results in the classification is also detailed. In addition, an explanation of the architecture and training of the model presented in this chapter is provided, and in Figure 5.1, the flow chart of the algorithm used is presented. To obtain a reference on the performance of our model proposed in this work, a comparison is made with VGG-NET.

5.3.1 Dataset description

For the training of the model, a dataset has been created from the combination of databases from the RSNA and Kaggle, with a total of 4618 X-rays images. The dataset contains X-ray images of COVID-19, pneumonia and normal images of patients who do not present virus infection. These have been divided as follows: 3568 images were used for the training dataset, 525 images for the validation dataset and 525 for the test dataset. The distribution of the dataset for each of the three phases is illustrated in Table 5.1.

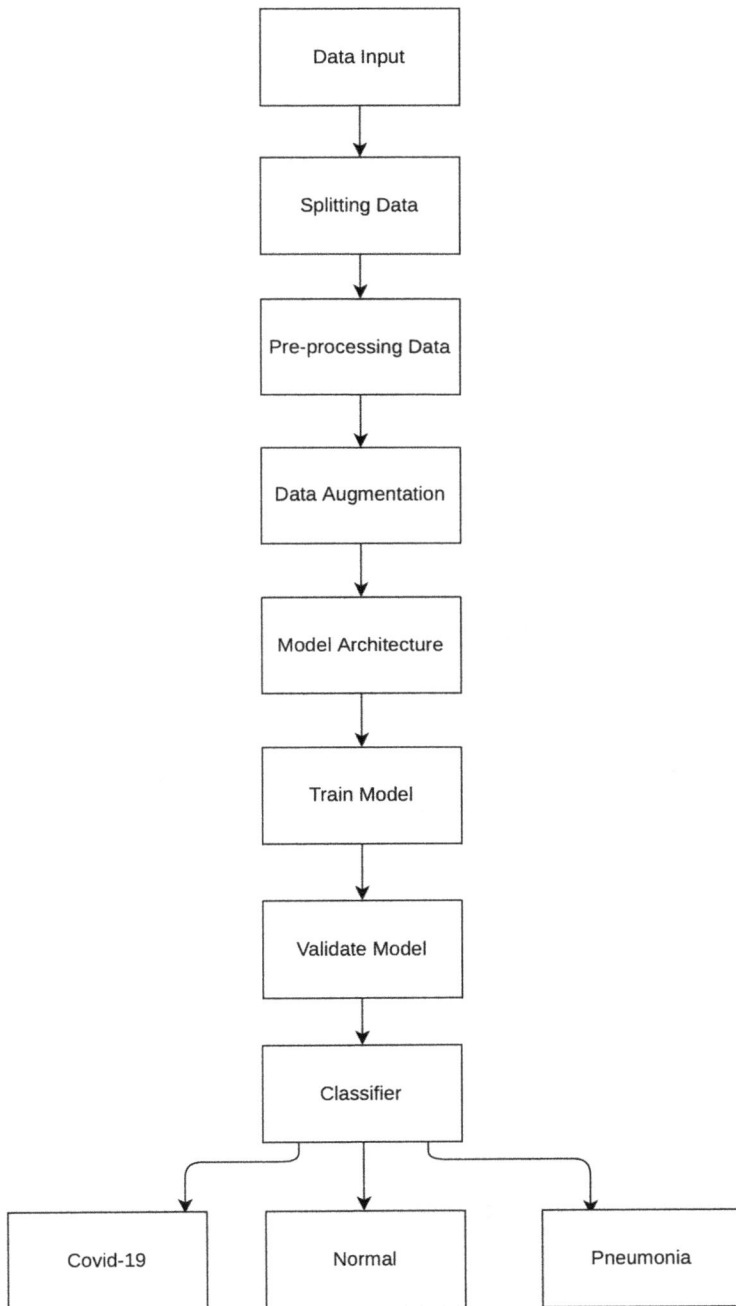

Figure 5.1 Flowchart for architecture proposal. Photograph by the authors.

Table 5.1 Dataset description

	COVID-19	Normal	Pneumonia
Train	1154	2770	694
Val	135	325	75
Test	135	325	75

5.3.2 Pre-processing

Prior to training, treatment was performed on the previously classified dataset. In this way, X-ray images with a single size of 128×128 pixels were implemented. In addition, a data augmentation was performed from the use of zoom range, the sheer range and horizontal flip properties with the Python ImageDataGenerator library, in order to generate better results in training as well as in model validation. Both the previous treatment of the data and the training and validation of the model were carried out using the Google Colab development environment, with the TensorFlow and Keras libraries under version 3.7.12 of Python.

5.3.3 Convolutional neuronal network

The present model has four convolutional layers with $64, 128, 256$ and 512 filters and a kernel size of 3. All these layers have "relu" as the activation function since it was the one that showed better performance compared to the function softmax. Along with these convolutional layers, there is a layer of BatchNormalization, MaxPooling and Dropout. At the outlet, there is a flattened layer together with two dense layers; this last layer has three outputs for Normal, Covid, Pneumonia. Figure 5.2 shows the architecture of the model proposed in this study.

5.4 EXPERIMENTAL RESULTS

In this section, the results obtained during the evaluation on the experimental configuration, the performance metrics and the type of artificial neural network used for each generated learning model are presented. Furthermore, the parameters obtained by each learning model are analyzed and compared in the selection of the most optimal. Finally, the results obtained are contrasted with those presented in the related works.

5.4.1 Experimental setup

Regarding the parameters implemented during the training of the artificial neural network, it had a total of 50 epochs with precision as a tracking

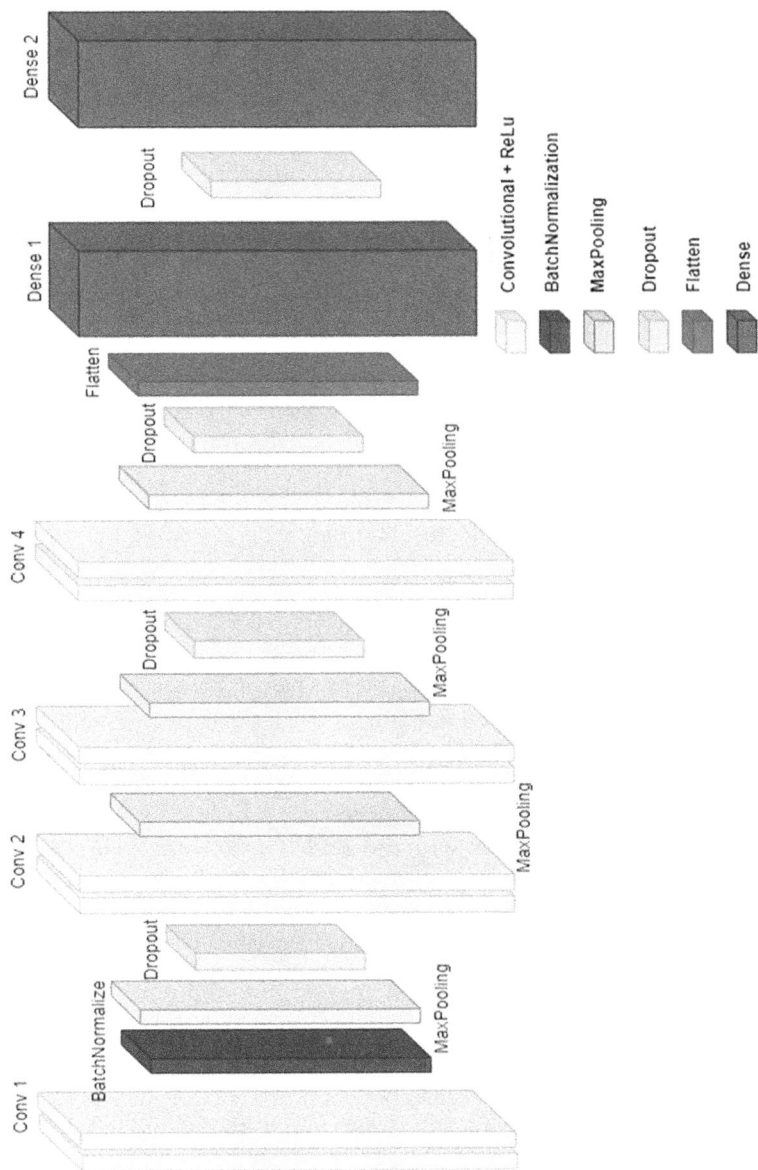

Figure 5.2 Descriptive diagram on the operation of the system. Photograph by the authors.

Table 5.2 Parameters of the proposed model

Class	Precision	Recall	F1-Score
COVID-19	0.93	0.93	0.93
Normal	0.96	0.97	0.96
Pneumonia	0.97	0.97	0.97

Table 5.3 Confusion matrix components

	Predicted positive	Predicted negative
Actual Positive	True Positive (TP)	False Negative (FN)
Actual Negative	False Positive (FP)	True Negative (TN)

metric and a learning index of 10X-4 under batches with a size of 32. The performance of the model was evaluated under the metrics of accuracy, accuracy and f1 score. The comparison of the data obtained can be better seen in Table 5.2.

After proceeding with the training of the model in batches with a size of 32 and 50 epochs, it was possible to obtain an accuracy of 96.86% for the training set and 95.42% for the validation set. The comparison between precision and loss for the training and validation sets is presented in Figure 5.3.

For the validation of the model, the set of 525 X-ray images was implemented and the results obtained were established in a confusion matrix where the true positives (TP), true negatives (TN), false positives (FP) and false negatives (FN) are presented in Figure 5.4. The structure of a confusion matrix is shown in Table 5.3. The average values for precision (0.953), F1-Score (0.953), recall (0.9566) are verified.

Finally, satisfactory results can be seen for the proposed three-class classification model. For comparison with the models used with VGG-Net, confusion matrices were created for each of the networks in the same way. In this way, it is easier to appreciate the performance of the model addressed in this book with other existing models. All the pieces of training were done using a device with a seventh-generation Intel Core i5 processor, with an NVIDIA GEFORCE 920Mx graphics card with 4GB of RAM memory; due to the low specifications, the training times were the maximum which enabled free use of the environment developer Google Colab.

Figure 5.5 shows the confusion matrix for the VGG-NET network, which had training of 60 epochs. Using this approach, it is possible to overcome the difficulty in detecting COVID-19 in the test dataset previously generated. However, good results are evident for the detection of pneumonia, as well as normal radiographs.

The plot of the values of both Accuracy and Loss for the training and validation sets shows that the values of the validation set did not completely approach those obtained for our training set. This is noticeable in Figure 5.6.

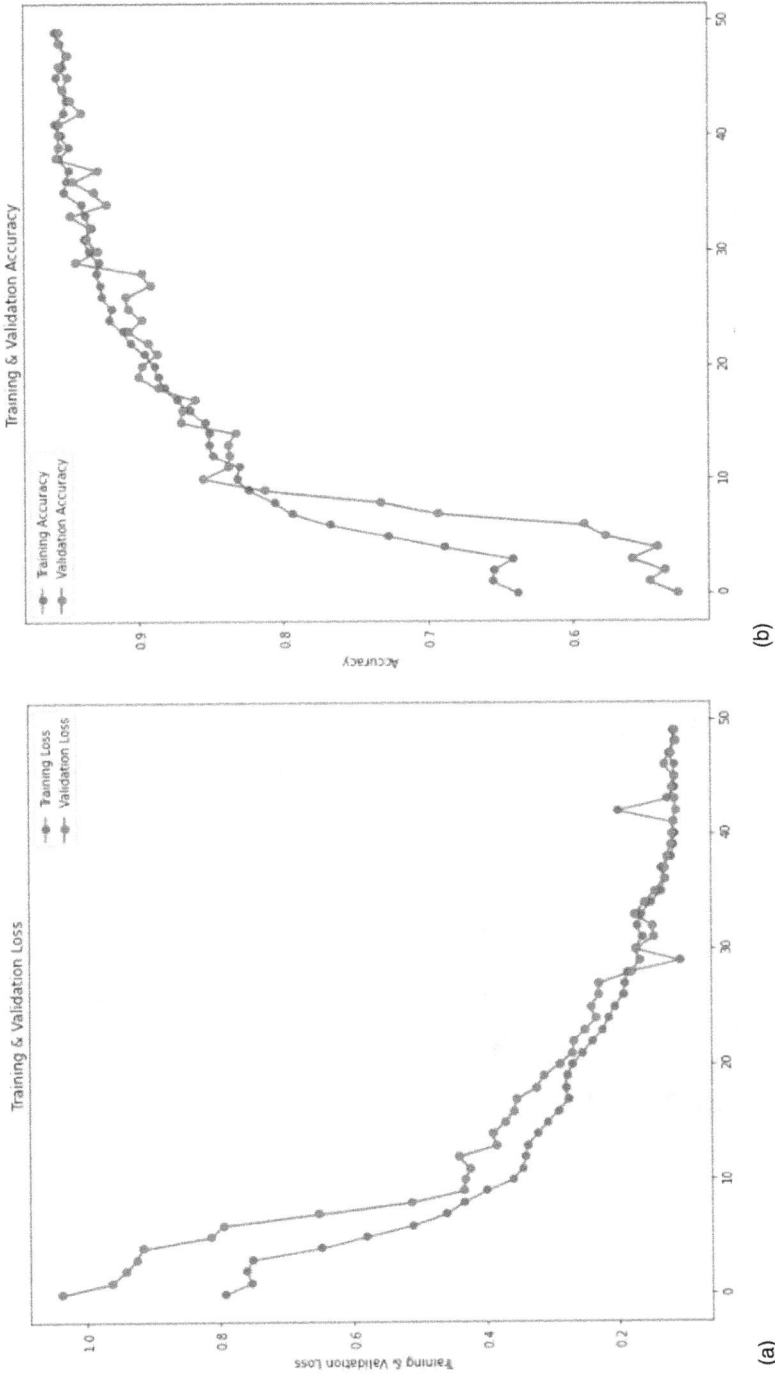

Figure 5.3 Training Accuracy-Loss versus epochs for the proposed Covid model. Photograph by the authors.

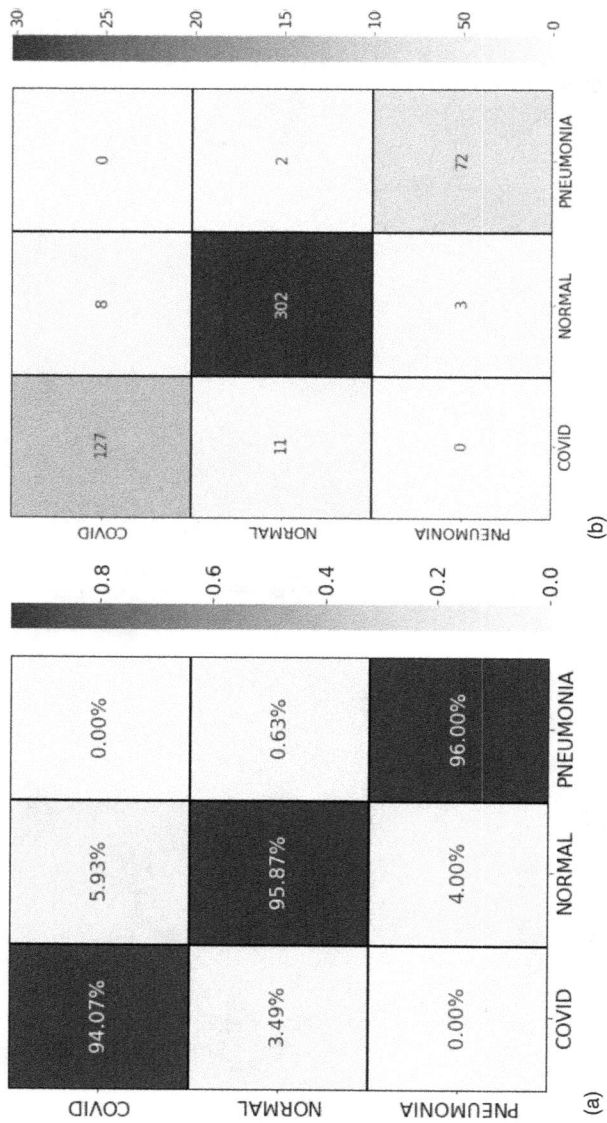

Figure 5.4 Confusion matrix with percentage and numbers of predictions. Photograph by the authors.

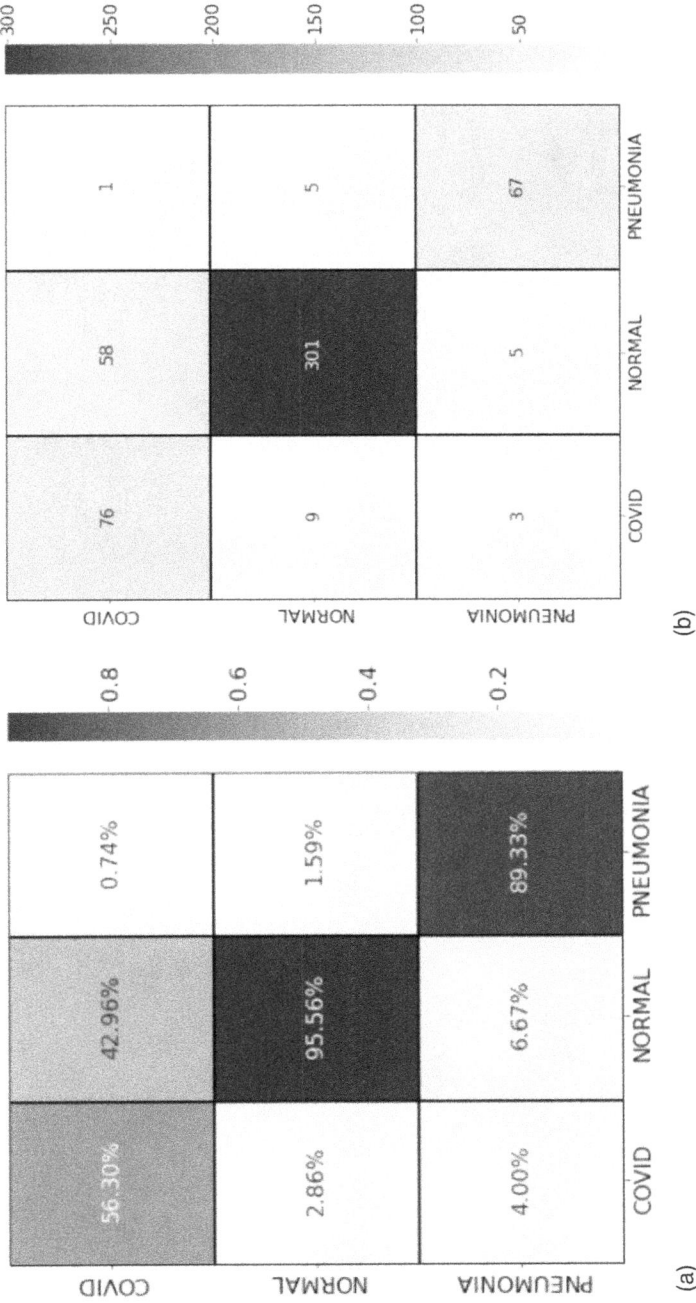

Figure 5.5 Confusion matrix with VGG-NET. Photograph by the authors.

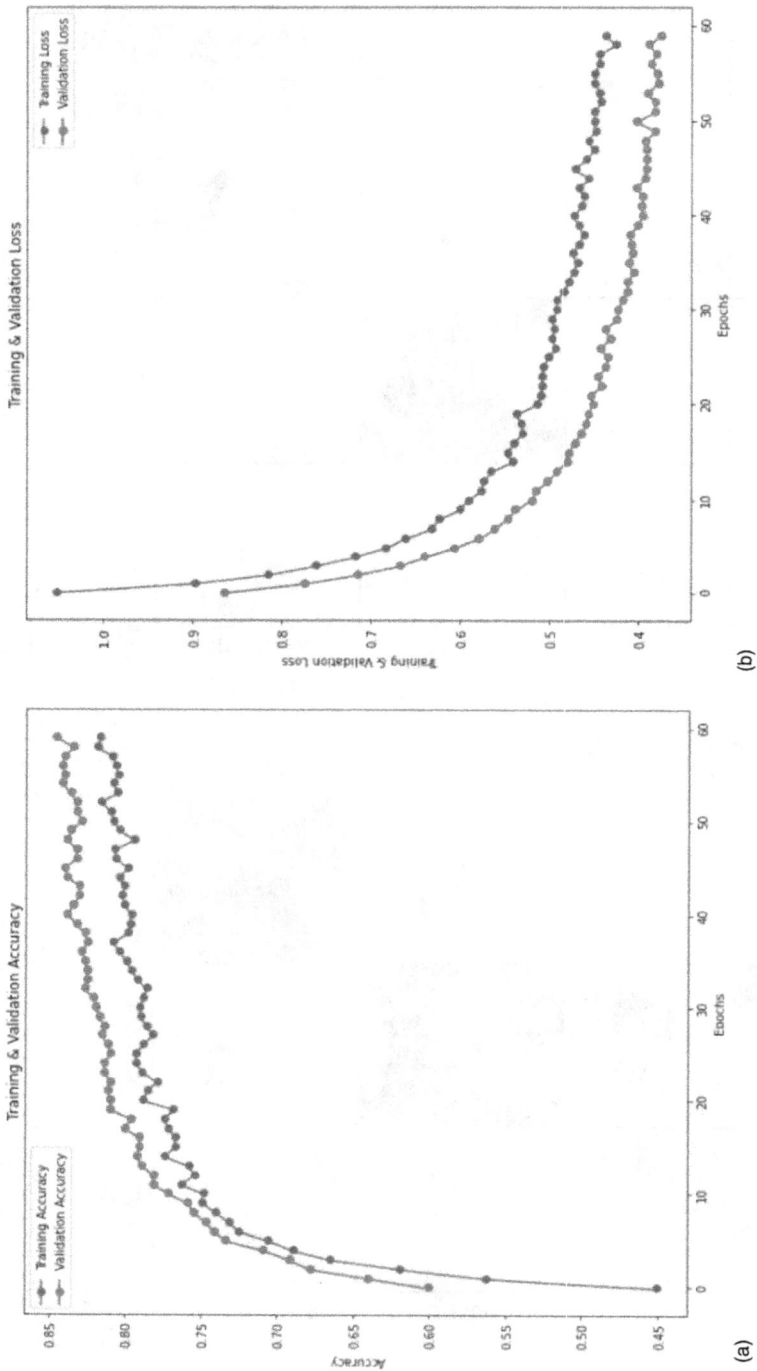

Figure 5.6 Training Accuracy-Loss versus epochs for VGG-NET. Photograph by the authors.

5.4.2 Data augmentation

In order to improve the results of the trained neural model, data augmentation techniques are used to increase the accuracy of the deep learning algorithm, given the difficulty in obtaining large volumes of data for the training of deep learning models and how unbalanced these sets are. It is necessary to perform this data treatment before training. In this study, 2770 normal images were used, which contains only 694 Pneumonia images. However, the use of data augmentation helps to reduce this difference. For this, the available Python libraries were used in this process.

5.4.3 Evaluation of performance metrics

For the evaluation of the metrics, four equations are needed to calculate the precision (5.1), recall (5.2), accuracy (5.3) and F1-Score (5.4) for the operation of the proposed neural network model, which can be defined as follows.

$$\text{Precision} = \text{TP} / (\text{TP} + \text{FP}) \times 100\% \tag{5.1}$$

$$\text{Recall} = \text{TP} / (\text{TP} + \text{FN}) \times 100\% \tag{5.2}$$

$$\text{Accuracy} = (\text{TP} + \text{TN}) / (\text{TP} + \text{FN} + \text{TN} + \text{FP}) \times 100\% \tag{5.3}$$

$$\text{F1} - \text{Score} = (\text{Precision} \times \text{Recall}) / (\text{Precision} + \text{Recall}) \tag{5.4}$$

5.4.4 Graphical user interface

Regarding the design of the graphical interface for the desktop, it was developed in the Python environment, implementing the Tkinter library oriented to the development of graphical user interfaces. In this, the user is asked to search for the file to be analyzed to later indicate if the diagnosis corresponds to one of the three available categories: COVID-19, Pneumonia or Normal described in Figure 5.7. Additionally, it provides the number of registered cases since the application is executed. A graphical representation of the statistics generated in the mobile application is illustrated in Figure 5.8.

In addition, it is convenient to indicate that the learning models developed in Python are being modified to be executed on mobile devices with limited processing and storage capacities. This enables diagnosis locally and without internet connection in hospital centers through libraries for TensorFlow purposes, greatly reducing waiting times, costs in specialized equipment and exposure to the contagion of the disease by health personnel.

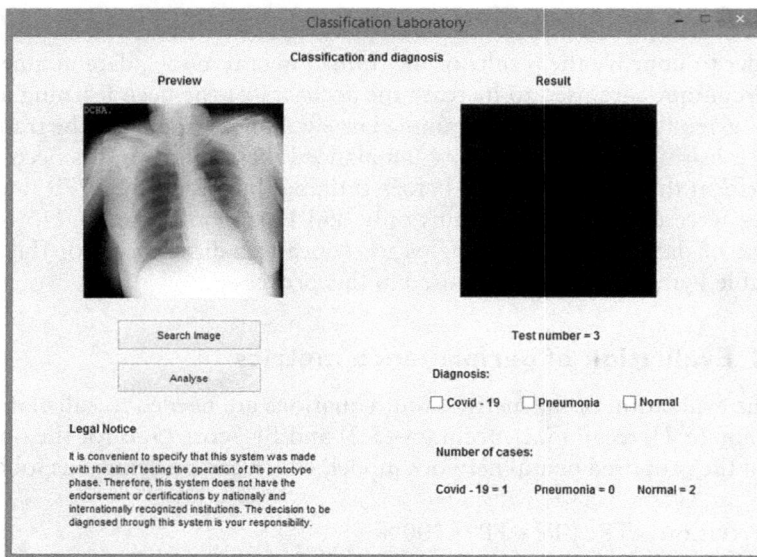

Figure 5.7 Graphical user interface for desktop devices. Photograph by the authors.

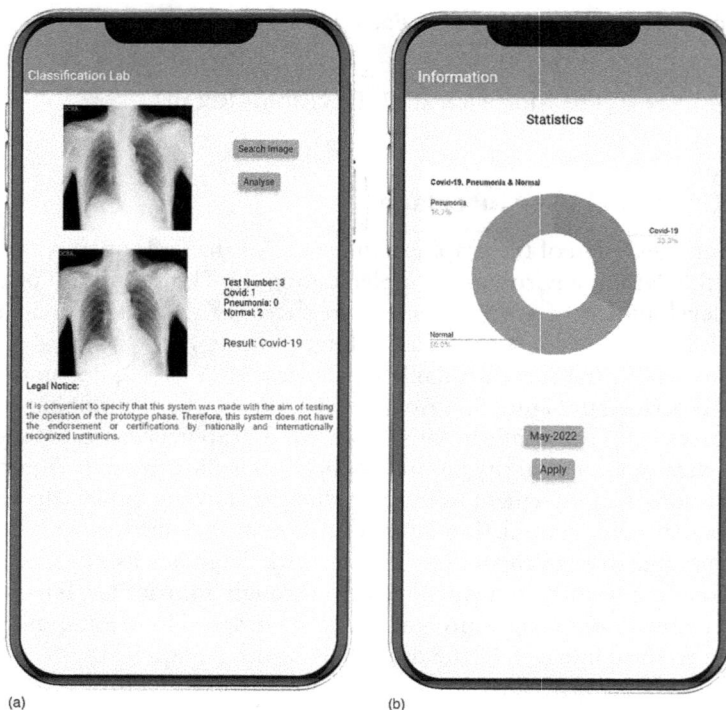

(a)   (b)

Figure 5.8 Graphical user interface for mobile devices. Photograph by the authors.

5.5 CONCLUSION AND FUTURE WORK

A low-cost method that enables automatic detection of COVID-19 and pneumonia in patients has been presented using deep learning techniques with the aim of preventing the spread of these diseases to other people. From this, an initial indicator is developed to establish the guidelines for the prevention of contagion through isolation or quarantine measures until an exhaustive study significantly limits the degree of exposure of the personal doctor in these cases.

The deep learning neural model allowed assertive classification and recognition of the test datasets entered during the sessions for performance validation. Based on the above, it was possible to visualize the significant effects in the automatic detection and extraction of characteristics associated with the diagnosis of COVID-19 and pneumonia, where an accuracy of 96.86% for the training set and 95.42% for the validation set, respectively, was obtained for a set of 4610 images of chest radiographs extracted from the Radiological Society of North America (RSNA) and Kaggle databases.

From a computational perspective, it is essential to enrich the collection of examples available in public databases with samples of COVID-19 and pneumonia. This is due to the fact that the performance of the classification and segmentation of images depends on the architecture and parameters entered in the network, and a more extensive radiography database would require greater performance in terms of the processing and storage capacity of the computer system. Therefore, the strategy proposed in this document may be subject to implications of accuracy, sensitivity and specificity in the datasets tested. Finally, as future work, it is intended to apply advanced image segmentation techniques in lung regions for the diagnosis of pneumonia in early stages in extensive datasets that allow emulating real scenarios and future applications.

REFERENCES

1. Panwar, H., Gupta, P. K., Siddiqui, M. K., Morales-Menendez, R., & Singh, V. (2020). Application of deep learning for fast detection of COVID-19 in X-Rays using nCOVnet. *Chaos, Solitons & Fractals*, 138, 109944.
2. Ozturk, T., Talo, M., Yildirim, E. A., Baloglu, U. B., Yildirim, O., & Acharya, U. R. (2020). Automated detection of COVID-19 cases using deep neural networks with X-ray images. *Computers in Biology and Medicine*, 121, 103792.
3. Pathak, Y., Shukla, P. K., Tiwari, A., Stalin, S., & Singh, S. (2020). Deep transfer learning based classification model for COVID-19 disease. *IRBM*, 1, 112–119.
4. Khan, A. I., Shah, J. L., & Bhat, M. M. (2020). CoroNet: A deep neural network for detection and diagnosis of COVID-19 from chest X-ray images. *Computer Methods and Programs in Biomedicine*, 196, 105581.
5. Brunese, L., Mercaldo, F., Reginelli, A., & Santone, A. (2020). Explainable deep learning for pulmonary disease and coronavirus COVID-19 detection from X-rays. *Computer Methods and Programs in Biomedicine*, 196, 105608.

6. Nayak, S. R., Nayak, D. R., Sinha, U., Arora, V., & Pachori, R. B. (2021). Application of deep learning techniques for detection of COVID-19 cases using chest X-ray images: A comprehensive study. *Biomedical Signal Processing and Control*, 64, 102365.

7. Belman-López, C. E. (2022). Detection of COVID-19 and other pneumonia cases using convolutional neural networks and X-ray images. *Ingeniería e Investigación*, 42(1), 164–173.

8. Mahmud, T., Rahman, M. A., & Fattah, S. A. (2020). CovXNet: A multi-dilation convolutional neural network for automatic COVID-19 and other pneumonia detection from chest X-ray images with transferable multi-receptive feature optimization. *Computers in Biology and Medicine*, 122, 103869.

9. Marquioni, V. M., & De Aguiar, M. A. (2020). Quantifying the effects of quarantine using an IBM SEIR model on scalefree networks. *Chaos, Solitons & Fractals*, 138, 109999.

10. Catal Reis, H. (2022). COVID-19 diagnosis with deep learning. *Ingeniería e Investigación*, 42(1), 253–261.

11. Greenspan, H., Van Ginneken, B., & Summers, R. M. (2016). Guest editorial deep learning in medical imaging: Overview and future promise of an exciting new technique. *IEEE Transactions on Medical Imaging*, 35(5), 1153–1159.

12. Karar, M. E., Hemdan, E. E. D., & Shouman, M. A. (2021). Cascaded deep learning classifiers for computer-aided diagnosis of COVID-19 and pneumonia diseases in X-ray scans. *Complex & Intelligent Systems*, 7(1), 235–247.

13. Amyar, A., Modzelewski, R., Li, H., & Ruan, S. (2020). Multi-task deep learning based CT imaging analysis for COVID-19 pneumonia: Classification and segmentation. *Computers in Biology and Medicine*, 126, 104037.

14. Shi, F., Wang, J., Shi, J., Wu, Z., Wang, Q., Tang, Z., ... & Shen, D. (2020). Review of artificial intelligence techniques in imaging data acquisition, segmentation, and diagnosis for COVID-19. *IEEE Reviews in Biomedical Engineering*, 14, 4–15.

15. Pham, D. L., Xu, C., & Prince, J. L. (2000). A survey of current methods in medical image segmentation. *Annual Review of Biomedical Engineering*, 2(3), 315–337.

16. Badrinarayanan, V., Kendall, A., & Cipolla, R. (2017). Segnet: A deep convolutional encoder-decoder architecture for image segmentation. *IEEE Transactions on Pattern Analysis and Machine Intelligence*, 39(12), 2481–2495.

17. Vieira, P., Sousa, O., Magalhães, D., Rabêlo, R., & Silva, R. (2021). Detecting pulmonary diseases using deep features in X-ray images. *Pattern Recognition*, 119, 108081.

18. Das, N. N., Kumar, N., Kaur, M., Kumar, V., & Singh, D. (2020). Automated deep transfer learning-based approach for detection of COVID-19 infection in chest X-rays. *IRBM*, 1, 334–343.

19. Cortés, E., & Sánchez, S. (2021). Deep learning transfer with AlexNet for chest X-ray COVID-19 recognition. *IEEE Latin America Transactions*, 19(6), 944–951.

20. Basavegowda, H. S., & Dagnew, G. (2020). Deep learning approach for microarray cancer data classification. *CAAI Transactions on Intelligence Technology*, 5(1), 22–33.

21. Kwon, O., Yong, T. H., Kang, S. R., Kim, J. E., Huh, K. H., Heo, M. S., ... & Yi, W. J. (2020). Automatic diagnosis for cysts and tumors of both jaws on panoramic radiographs using a deep convolution neural network. *Dentomaxillofacial Radiology*, 49(8), 20200185.
22. Manickam, A., Jiang, J., Zhou, Y., Sagar, A., Soundrapandiyan, R., & Samuel, R. D. J. (2021). Automated pneumonia detection on chest X-ray images: A deep learning approach with different optimizers and transfer learning architectures. *Measurement*, 184, 109953.
23. Abdelkader, R., Ramou, N., Khorchef, M., Chetih, N., & Boutiche, Y. (2021). Segmentation of X-ray image for welding defects detection using an improved Chan-Vese model. *Materials Today: Proceedings*, 42, 2963–2967.
24. Pang, K., Ai, D., Fang, H., Fan, J., Song, H., & Yang, J. (2021). Stenosis-DetNet: Sequence consistency-based stenosis detection for X-ray coronary angiography. *Computerized Medical Imaging and Graphics*, 89, 101900.

Chapter 6

Fuzzy logic and applications

Raabia Kausar
Lovely Professional University, Phagwara, India

Rukia Rahman
University of Kashmir, Srinagar, India

Yogeta Pimpale
Lovely Professional University, Phagwara, India

CONTENTS

6.1 INTRODUCTION

Fuzzy logic is defined as a method of reasoning that uses several values rather than the binary options true or false and is derived from fuzzy set theory. Numerous fuzzy logic variables have truth values that range from 0 to 1. It is an extension of fuzzy set theory, where inferences are more illogical than precise. Fuzzy logic places a high focus on natural language. Fuzzy set theory is the main method used to achieve fuzzy logic's ultimate goal, which is to provide a solid foundation for approximation reasoning with uncertain propositions. In order to imitate the uncertainty of natural language, Dr. Lotfi Zadeh, a professor at the University of California at Berkley, first devised it in the 1960s. According to him, we should view the process of "fuzzification" as a methodology to extend every particular theory from discrete to a continuous (fuzzy) configuration rather than thinking of fuzzy theory as a single theory. The main goal of fuzzy logic-based structures is to mimic being natural in handling and resolving issues that could not totally be codified via the use of analytical models and handled through the application of system theory methods [1]. In fuzzy process control, knowledge of human operating criteria, process conditions, and input-output linkages is encoded into a system in the form of language descriptions. Fuzzy inference actions are used to encode the control activities.

The foundation of fuzzy logic is an anthropoid thought and everyday actions. It can be put into practice using technology, computer-aided learning, or a blend of the two. It can be integrated into a diverse range of products, from miniature portable equipment to immense automated systems for controlling the processes. Automotive manufacturers are using the fuzzy logic system in the current competitive environment to improve quality, cost, and development time reduction. Although fuzzy logic was initially intended to be a better way of managing and organizing data, it has turned out to be a great option for many industrial automation applications. Additionally shown are the potential applications of fuzzy control systems to RACs in the future.

6.2 FUZZY LOGIC CONCEPTS

Fuzzy control methods have drawn a lot of attention and are now crucial to contemporary control engineering. A fuzzy system can serve as a universal approximator to nonlinear functions, thanks to the usage of linguistic knowledge in the form of IF-THEN rules. A logical system called fuzzy logic is an expansion of multi-valved logic. Multi-valved logic in a logics system is a propositional calculus with more than two truth values [2].

The following ideas are usable with fuzzy logic:

Fuzzy Predicate: In fuzzy logic, the fuzzy predicate $F(x)$ represents the fuzzy statement x is F, where F is a fuzzy number. Multivalued truth values, such as "reasonably true," "very true," "fairly false," and "false," among others, are used in fuzzy logic.

Fuzzy quantifiers: The term "fuzzy quantifier" is used to describe fuzzy numbers that participate in or are a component of fuzzy propositions. Fuzzy quantifiers defined on real numbers are one of its varieties. Fuzzy quantifiers with a defined 0 to 1 interval.

6.3 REASONS FOR USING FUZZY LOGIC SYSTEM

Fuzzy logic is adaptable and intuitive technically.
Fuzzy logic can handle imperfect data.
Fuzzy logic is capable of simulating nonlinearities of any intricacy.
It can be used in conjunction with conventional control techniques.
Natural language is the foundation of fuzzy logic control.

6.3.1 Main usage of fuzzy logic systems is prevalent in the following areas

Fuzzy logic approaches have been applied in various household goods, including televisions, microwaves, washing machines, and other appliances, to progress the technology. Figure 6.1 shows the "principle" of Fuzzy Logic and its different applications.

6.3.2 Fuzzy logic in washing machines

Washing machines using fuzzy logic are becoming more common. Efficiency, effectiveness, ease of use, and lower costs are all benefits that these machines

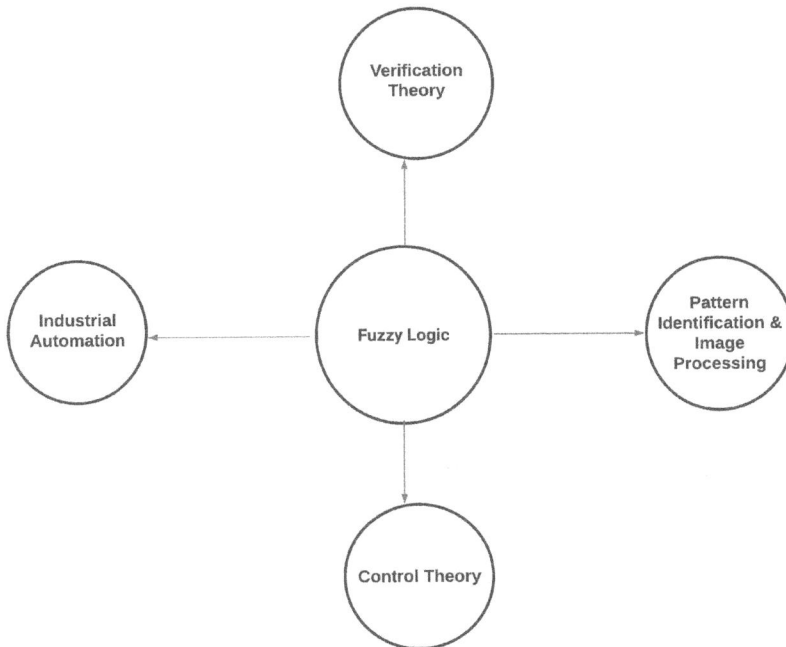

Figure 6.1 Different fields of fuzzy logic.

provide. Fuzzy logic-based washing machines can detect the measure of filth, the quantity of detergent, amount of water to add, and so on. Sensors regularly monitor the machine's changing internal conditions and modify operations as necessary to achieve the greatest wash outcomes. The washing process, water intake, water temperature, wash duration, rinse effectiveness, and spin speed are all controlled by fuzzy logic. By doing this, the lifespan of a washing machine is enhanced. The washer's load is reevaluated by the machine to guarantee proper spinning. Otherwise, if an unbalance is found, it slows down the spinning speed. Optical sensors are used by neurofuzzy technology to detect filth in the water, and a textile sensor determines the kind of textile and adjusts the wash cycle appropriately.

6.4 IMAGE PROCESSING

The field of fuzzy image processing encompasses many methods that comprehend, represent, and treat pictures, image segments, and feature sets as fuzzy sets. The chosen fuzzy technique and the issue at hand determine how the representation and processing are done. As shown in Figure 6.3, there are three basic phases of fuzzy image processing: image fuzzification,

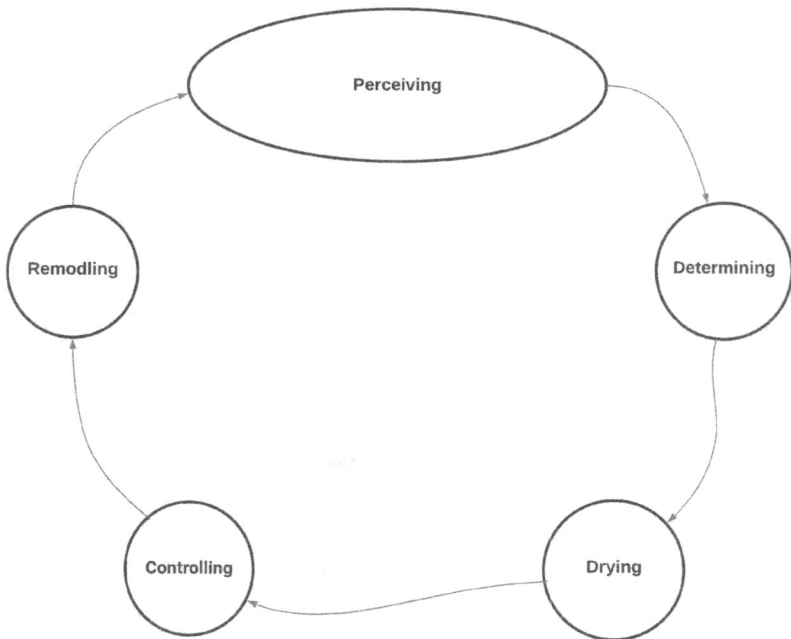

Figure 6.2 Fuzzy logic in a washing machine.

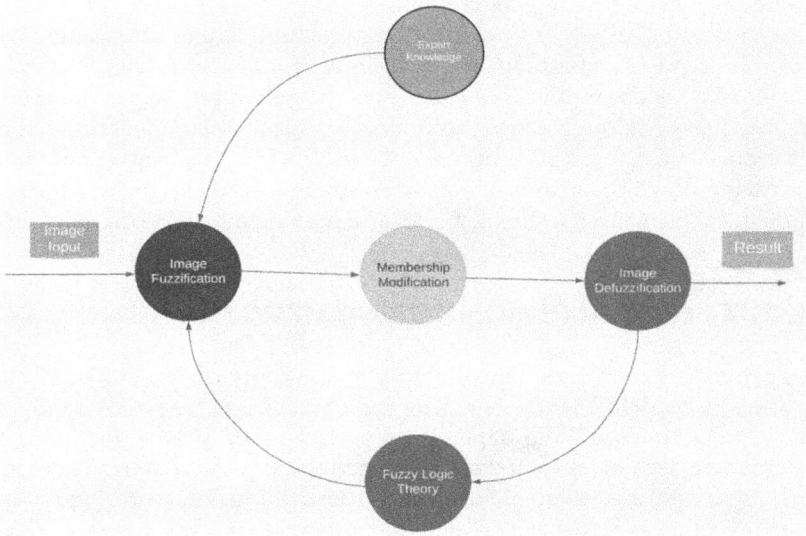

Figure 6.3 Structure of image processing.

membership value adaptation, and image defuzzification. Since we lack fuzzy hardware, the fuzzification and defuzzification processes are necessary. The encoding of picture data (fuzzification) and the explication of the final outcome are hence the processes which allow the analysis of images using fuzzy approaches (defuzzification). Fuzzy image processing excels in the midpoint, which requires modifying the membership values. Once the picture data has been fuzzified from the gray-level plane to the membership plane, the membership values are updated using the proper fuzzy techniques. This can be done using either a fuzzy regime-based technique or a fuzzy integration accession.

6.5 ANESTHESIA DEPTH CONTROL USING FUZZY LOGIC

To put patients to sleep throughout most surgical procedures, in hospitals, manual methods are used. The manual mechanisms are circumstances that are either ON or OFF. Manual systems do not have interval values between ON and OFF; therefore, anesthetic operations cannot be safe and relaxing. Fuzzy logic control is used to regulate anesthesia for such a reason. An objective method using fuzzy logic is presented in this chapter for administering anesthesia to patients undergoing surgery. The theory of fuzzy logic is a broad mathematical strategy that permits limited memberships. Numerous research studies have demonstrated that fuzzy logic control is an effective way of managing complicated systems. The fuzzy logic system inputs T and N, which are obtained from patients under anesthesia, stand in for blood pressure (mmHg) and pulse rate (bpm), respectively. Anesthesia Output (AO) is the output of a fuzzy logic system. Fuzzy logic control during anesthesia has the potential to improve patient safety and comfort, and direct anesthetist focus to other physiological variables; they must control by slowing down their tasks, use the best anesthetic agent, protect the environment by using an anesthetic agent, and lower surgical procedure costs.

6.6 FUZZY LOGIC IN MONITORING DRUG DOSAGE

Fuzzy logic is widely used in evaluating the quantity of drug to be given to a particular patient [3]. Determining the appropriate medicine dosage for each patient is crucial and depends on a number of factors, including the patient's age, weight, sex, medical history, blood type, and more. There is no method for figuring out medicine dosage based on these factors. Fuzzy logic-based approaches have been created and applied in many areas of medicine, from the detection of tumors to the treatment of asthma. For the treatment of chronic intestinal inflammation, one of the applications is calculating the drug dose using FES (Functional Electrical Simulation). For the creation of FES, symptoms of the condition, such as deposition and prostate-specific antigen, were utilized to calculate the dosage of the medication salazopyrine. More than ten patients' data were used to determine the appropriate medication dosage for each patient. The outcomes of certain patients were compared to the dosages that their doctors had advised them to take. As a result, it has been seen that the suggested system has assisted doctors in reducing the length of therapy and has reduced bad effects associated with drug dosage calculation.

6.7 FUZZY LOGIC-BASED WATER TREATMENT SYSTEM OPTIMIZATION

The focus of this case study is a fuzzy logic approach utilized in biochemical creation, and the largest mouth penicillin production facility in the world is situated in Austria. Once penicillin has been removed from the bacteria that made it, the leftover biomass is treated using a wastewater treatment system. The fermentation sludge created following this treatment contains traces of microorganisms and nutritious salt. It serves as the basis for premium fertilizer and is a byproduct of the synthesis of penicillin. To create the fertilizer, the sludge is concentrated in a decanter and the leftover water is removed in a vaporizer. To bring down vitality expenditures related to the vaporizing process, the decanter's segregation of water and arid substance needs to be changed. Prior to using fuzzy logic, the operation was manually supervised by operators.

6.8 APPLICATIONS OF FUZZY LOGIC IN INDUSTRIAL AUTOMATION

Fuzzy logic has recently demonstrated its extensive future in industrial automation usance. Engineering professionals mainly rely on tried-and-true ideas in this application field. They primarily utilize ladder logic, a programming language used on programmable logic controllers, for discrete event control. Ladder logic is similar to electrical wiring schemes (PLC). Bang-bang type or Proportional-Integral-Derivative (PID) type controllers are typically used for continuous control. PID type controllers function effectively when the process they are controlling is steady, but they struggle in other situations:

A) The existence of significant disruptions (nonlinearity)
B) Process parameters that change over time (nonlinearity)

To regulate the water bath's temperature, a fuzzy logic controller (FLC) was created [4]. The FLC was found to respond more quickly than the traditional PID controller. Additionally, it was found that FLCs are far more similar in spirit to human reasoning and decision-making. Since building a mathematical model of the system is challenging due to limited understanding of the system, FLC can nevertheless be constructed.

6.9 SOFTWARE ENGINEERING

Any software or application development is a very difficult and vital task. A key significance for fuzzy logic in the software engineering is to address issues with the software development process [5]. Ahlawat et al. developed a method in 2015 for calculating the expenses and manpower involved in the software development process. They combine fuzzy logic and the Constructive Cost Model (COCOMO). The examined study illustrates how to use the provided model and fuzzy logic to estimate the scope, effort, and cost factors of any project. When building software, four classifiers are used to assess the process involved.

6.10 POWER SYSTEM APPLICATIONS

Controls for generating units include excitation control and prime mover control with automatic voltage regulation (AVR) and power system stability (PSS). The first control regulates variations in the energy supply system's generator speed as well as variables like boiler pressure or water flow. By using excitation control, the generator terminal voltage and reactive power output are kept within their machine-dependent ranges. System generation control is used to choose the active power output so that the system's overall generation can keep up with the system load. Additionally, it regulates the frequency and tie line flow between various power system sectors. Transmission control also keeps an eye on equipment for controlling power and voltage, such as tap-changing transformers, synchronous condensers, and static VAR compensators. In fact, every control has an impact on both parts and systems. A well-tuned PSS, for instance, can effectively offset the local mode oscillations and inter-area oscillations that the AVR is known to induce.

6.11 FUZZY CONTROLLERS' POTENTIAL FOR USE IN RACS IN THE FUTURE

Vapor pressure refrigeration systems are extensively employed in a variety of areas, notably industrial, commercial, and residential. In fact, the demand for these refrigeration and air conditioning systems, or RACs, has surged dramatically in recent years. The one-on-one implementation of this form of regulator marked beginning of fuzzy regulators' evolution as an application to RACs. In order to build real systems, fuzzy controllers have been employed over the years in conjunction with conventional controllers and cutting-edge control systems. These controllers are built to keep the ideal temperatures constant while consuming less energy. As a result, it is planned that RACs will utilize energy resources more effectively than current systems in the future, making them more energy efficient than those systems. Due to their considerable role in the global consumption of electricity, RACs represent a promising market in this regard [6]. The primary benefit of fuzzy controllers is their ability to simultaneously incorporate a huge number of variables and rules, allowing them to make decisions about how to use the system based on a set of applicable thermic conditions and constraints. Fuzzy controllers can comply with more fast controllers; therefore, it is anticipated that refrigerators and air conditioning systems will be able to predict operating situations in the future while maintaining acceptable thermal conditions, environment-friendly, and more effectively regulating energy usage.

Numerous applications are available in numerous disciplines as well. Some of them are as follows:

- Fuzzy knowledge-based system for managing the burning of waste in an incinerator facility.
- Using fuzzy control to ensure that sophisticated refrigeration systems operate at their best.
- A Fuzzy Expert System Architecture for Autonomous Small Satellite Avionics Planning and Scheduling.
- Fuzzy Antilock Brake System with Fuzzy Logic-Based Vehicle Speed Estimation.

- Using language filters in chat rooms and message boards or removing inappropriate text.
- Washers, dryers, rice cookers, and other home appliances.
- Escalators, Projectors, webcam, and computer network designs.

6.12 THE BENEFITS OF FUZZY LOGIC

Following are the benefits of fuzzy logic:

- Conceptually, fuzzy logic is simple to comprehend.
- It is flexible; adding extra functionality to a system is easy.
- In comparison to traditional approaches, fuzzy logic requires less time to develop.
- It is frequently employed in all spheres of life and gives simple, efficient responses to issues of tremendous complexity.

Because the techniques may be simply represented with little data, it does not require a large storage.

6.13 DISADVANTAGES OF FUZZY LOGIC

There is no formal, structured approach to building fuzzy systems. Only when they are basic, they are understandable.

Fuzzy logic systems require extensive assessment for both validation and verification.

Fuzzy logic systems have a lengthy run time and produce results slowly.

6.14 CONCLUSION

As a result of Lotfi Zadeh's work on the theory of fuzzy sets, the term "fuzzy logic" came into use. An entirely different, unconventional method of approaching a control problem is offered by fuzzy logic. Instead of attempting to comprehend how the system functions, this approach focuses on what it should accomplish. In order to process input data more akin to a human operator, it employs an imprecise but highly detailed language. It frequently works right away after implementation with little to no adjusting because of exceedingly durable and tolerance of function and information input. Instead of attempting to mathematically model the system, if that is even possible, one can concentrate on finding a solution for the issue. This almost always results in speedier, more affordable solutions. Fuzzy logic is thus a useful instrument with uses in the areas of processing and materials technologies that can promote creativity in research and industry. Once implemented, such techniques are simple to use, and the outcomes are usually pleasant and surprising. In recent years, fuzzy logic applications have grown in both diversity and number. Applications include industrial process control, medical instrumentation, portfolio selection, and consumer goods, including webcams, video recorders, washers and dryers, and kitchen appliances.

REFERENCES

1. J. Ravikumar, "Fuzzy Logic Control System and Its Applications", 08, 06, 01, n.d.
2. K. Mahesh Prasanna, C. Shantharama Rai, "Applications of Fuzzy Logic in Image Processing – A Brief Study", *Compusoft*, 05, 03, 1556, 2015.
3. P. Singh, K. Arora, U. C. Rathore, "Control Strategies for Improvement of Power Quality in Grid Connected Variable Speed WECS with DFIG–An Overview", *Journal of Physics: Conference Series*, 1, 2087–2093, 2022.
4. Y. Misra, Prof. (Dr.) H. R. Kamath, "A Review on Application of Fuzzy Logic in Increasing the Efficiency of Industrial Process", *International Journal of Latest Trends in Engineering and Technology*, 01, 03, 112, 2012.
5. R. Kaur, A. Singh, "Fuzzy Logic: 'An Overview of Different Application Areas'", *Advances and Applications in Mathematical Sciences*, 18, 08, 10, 2019.
6. J. M. Belman-Flores, D. A. Rodríguez-Valderrama, et al. "A Review on Applications of Fuzzy Logic Control for Refrigeration Systems", *Applied Sciences*, 12, 1302, 2022.

Chapter 7

IoT-based smart monitoring topologies for energy-efficient smart buildings

R. Senthil Kumar, P. Lenin Pugalhanthi, N. Dhanyaa,
V. Kaviyanjali, K. Dinesh and K. Kathirvel
Sri Krishna College of Technology, Coimbatore, India

CONTENTS

DOI: 10.1201/9781003407409-7

7.1 INTRODUCTION

The field of information and communication technologies is developing to a great extent nowadays due to its application in various types of sensors and advanced communication techniques. It enhances wireless communication that provides better connectivity and accessibility [1]. The Internet of Things (IoT) is an integrated network of sensors, software, and n number of technologies to exchange and connect to anything from anywhere to anyplace at anytime [2]. IoT works by connecting sensors, edge devices, and cloud, in which the sensors gather the data and send them to the cloud by an edge device or a gateway to examine and analyse the data [3]. These processes are carried out without any human interventions, although people can monitor and interact with the devices from anywhere.

IoT system also depends upon other technologies like artificial intelligence (AI), data optimization techniques and machine learning (ML) to provide a much easier and safer way of collecting data and information at a dynamic speed [4]. IoT helps improve the lives of people and business environment by offering smart devices like automated homes, vehicles and devices, which were used everyday. IoT enhances the performance of the logistics operations and systems to provide improved security service with cheaper costs. The chances of hacking the information and data are larger when connected to a number of devices. This is the great disadvantage of the system, which turns into an advantage of hackers to steal confidential information. Also, when there is a bug in a single device, it affects other connected devices and gets corrupted.

Many researchers are working on these issues to improve the security, privacy and connectivity concerns, which were suggested by the people. IoT networks mark their excellence in many domains like healthcare, smart cities, smart buildings, wearables devices, agriculture and so on. In the field of smart buildings, IoT networks reduce energy costs by using sensors to provide correct temperature to the room according to the number of persons. These processes can be monitored by people and can be done automatically [5]. Researchers and scientists are working on various techniques and sensors to make the smart home system more successful and easier to all the people. Earlier traditional methods were used to obtain energy management data through statistical analysis that were obtained using an energy meter [6]. But traditional methods do not have more accurate and hourly consumption data. By integrating various techniques, it is possible to obtain the hourly energy consumption rate, which is more useful to consume and avoid the wastage of energy, which was obtained by our precious resources. Energy prediction techniques based on hourly, weekly, monthly and yearly analysis are used with various parameters. Scientists and research scholars suggest that energy consumption should be followed by people; if there is no energy consumption, definitely, there will be a shortage of energy in future.

There are two options for energy:

1) Inventing various techniques to produce more amount of energy.
2) Following energy consumption methods to preserve already available energy resources.

These two criteria can be fulfilled by IoT-controlled smart homes. Inventing various techniques to produce more amount of energy is more expensive than the energy consumption technique, as it requires more capital investment and lot of time. On the other hand, the solution can be achieved by taking energy consumption measures. The first step is to predict the energy level to undergo the optimization process. The energy optimization technique is much helpful in maintaining the correct level of energy to different appliances, though each and every appliance consumes the same amount of energy every time. This is the reason for consuming more amount of energy in our houses [7].

Nowadays, various sensors have been designed to monitor and analyse the energy level in appliances. Also, a device can be automatically changed according to the various conditions and temperature range. These sensors can be controlled by humans in a more easier way. These sensors work without any human interventions and have the ability to adjust under different climatic conditions and at various temperature ranges. This concept is more useful in smart homes to maintain the safer range of energy in a device [8]. The performance of the techniques is based on the temperature, air quality, air flow, illumination, humidity and the external conditions of a particular area. Different techniques and algorithms have been proposed by various research scholars to improve energy management techniques in smart buildings. Wireless sensor networks (WSNs) are now becoming fundamental and crucial devices in the world, which provide improved stability in energy management and are much easier to operate and monitor the process even in our mobile phones [9].

Wireless sensor networks also help to reduce the installation cost of various sensors that were used in the developed environment. There are several wireless sensor technologies that are arising in the modern era, which can be designed according to the features, applications, design, construction and strength. These techniques can be operated through a battery, which needs to be plugged into a power supply. The selection of the battery is the most crucial step. The battery should maintain a better operating range and should have good life cycles [10]. The battery should be of replaceable type and its sending power and receiving power should be maintained at the correct level. One of the most important things in constructing the IoT networks and smart buildings is a fire alarm system, which is the essential thing in the smart building to ensure people's safety, lives and surrounding environment. Artificial intelligence helps to frame various algorithms; using these algorithms, IOT networks start to perform an operation. Integration of ML and big data helps to determine and identify the energy usage status, its availability, its shortage ratio and so on. IoT has the ability to develop a separate application for smart control of devices in the buildings, which can be done using a relay, microcontroller and a controller.

Many researchers are working in the area of smart buildings, smart sensors, WSNs, smart automation, energy consumption and so on. The primary goal is to improve current techniques and technologies. IoT in combination with AI provides an extraordinary behaviour in the field of decision-making and logistics [11]. Realistic approach has been carried out to implement smart building architecture, which was built using hardware and software levels. The hardware process consists of heterogeneous sensors, which are wireless and are used to sense the surrounding environment and collect the data [12]. The software process is divided into different stages like data logging and storage. In the software, the data and information were encoded, and it correlates the previous and current information to obtain the status of the system. The main aim of this chapter is to illustrate the characteristics, features and different monitoring techniques in smart buildings.

The chapter is structured as follows. Section 7.2 discusses the IoT-based smart building. Section 7.3 illustrates the architecture of the smart building. The monitoring techniques of smart buildings are extensively covered in Section 7.4. Section 7.5 discusses the opportunities and challenges of implementation of smart buildings. Section 7.6 concludes the chapter.

7.2 IOT-BASED SMART BUILDING

The majority of smart building's components are automated construction equipment and improved communication technologies. The systems are made up of an air conditioning system, a heating unit, an illumination source, sensors, window operators, elevators, air quality monitoring systems, and other electrical devices and programmes. Such specialised tools have been added to the platform of smart buildings to provide real-time monitoring and controls utilising cutting-edge technologies [13]. Based on the different communication protocols specified by their manufacturers, these systems are not directly connected to each other. To overcome this communication issue and to combine them on the same platform, there have been many trial-and-error issues to enhance the protocols involved in these systems.

An IoT architecture in a smart building is shown in Figure 7.1. A smart building can be totally controlled by the users and rapidly fulfil the demand of users [14]. The data involved in IoT systems consists of the individual needs of users and equipment, and indoor climatic conditions such as temperature, humidity and illuminance. Data is categorized and processed for controlling the building. The data generated are exposed as information about each transaction between an equipment and a user in the subsequent layer. Then, a time series will be created using this data [15].

Typically, IoT technology is used to connect various smart gadgets that make up smart buildings, including sensor devices, mobile phones and other smart devices. These devices have the ability to exchange the information and interact with each other [16]. There are many applications and services available through the IoT such as agriculture, healthcare, household

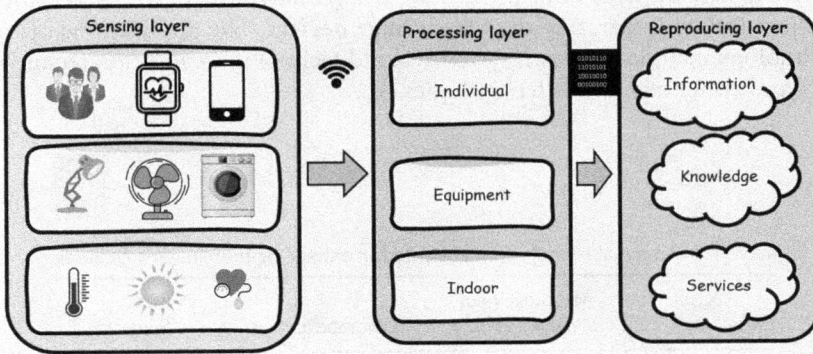

Figure 7.1 IoT architecture in a smart building.

appliances, military and so on. There are some upcoming projects based on IoT, which are ubiquitous connectivity, transporting system, delivering and some real-time applications [17]. An open-source intelligent system is also developed to allow monitoring and operations involved in smart home. The installed cameras and sensors are responsible for the data regarding light levels, resident behaviours and temperature. The data must be specified in a certain threshold; if it exceeds, then it can be noticed through a message/mail. To detect the typical situation, the system was developed by AI.

Thermal model of inhabitants for enhancing human care facilities is illustrated in Table 7.1. The advancement in machine learning, sensor devices and the IoT will give path to convert standard buildings into smart buildings [18]. By incorporating day lighting and improving the operation for occupancy detection and dimming, smart lighting reduces over lighting. Light level controls are developing quickly and becoming more and more popular across the industry. These smart-lightning systems can be operated wirelessly, which is monitored by lighting management systems. The wireless control is become easier by Retrofitting, whereas the user can handle smart lighting via web-based dash boards [19]. To adopt the remote monitoring system, the device should have the network provider with a larger area and very low power consumption.

7.2.1 Basic components of IoT

The IoT is actually a network, which is used to interconnect the devices that involves the functions such as share, communicate and to deliver the real-world data. The main theme of the smart building is to operate the devices wirelessly through internet; on that way, the IoT technology plays avital role to link the physical devices to the internet [20]. The block diagram of IoT integration in smart buildings is shown in Figure 7.2. These connections of device are made with built-in sensors, using analytics to give useful information and by integrating data from huge devices. The applications of IoT technology include detection systems, cloud technologies, location technologies and communications technologies.

Table 7.1 Thermal model of inhabitants for enhancing human care facilities

Class	Metabolic heat average (W/m²)	Metabolic heat	Example
Low	110	185	standing
Moderate	160	294	Sustained hand
High	220	417	Intense effort

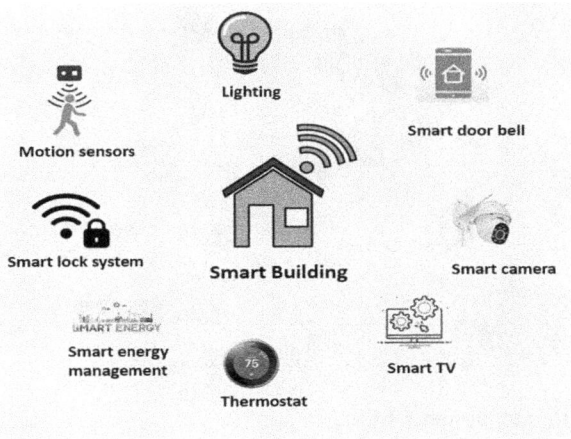

Figure 7.2 IoT integration in smart buildings.

7.2.2 Sensors

These IoT sensors are used to save all the important information on the sever and display it when needed [21]. By using these sensors, the users can be benefited and also enhance the environment of buildings.

7.2.3 Gateways

IoT gateways are majorly used for the communication purpose in the system with respect to interfaces, protocols and communication options [22].

7.2.4 Building monitoring systems

Building monitoring system is also referred to building automation system (BAS) because it plays an important role in energy management in industrial and commercial buildings [23]. There are many advantages by implementing the building monitor system such as reducing facility running cost, increasing safety, remote features and improving the convenience for facility managers.

7.2.5 IoT with integrating AI in smart buildings

The integration of AI and IoT in smart buildings will help the user to feel safe and comfortable. To reduce and increase the energy consumption and operational efficiency, respectively, the data from a range of sensors are utilized in IoT-based smart buildings [4]. The basic components of IoT in smart building are shown in Figure 7.3. The major role of IoT devices in smart buildings is to control and monitor the energy consumption. The technology of IoT is also used to identify the environmental impacts such as pressure,

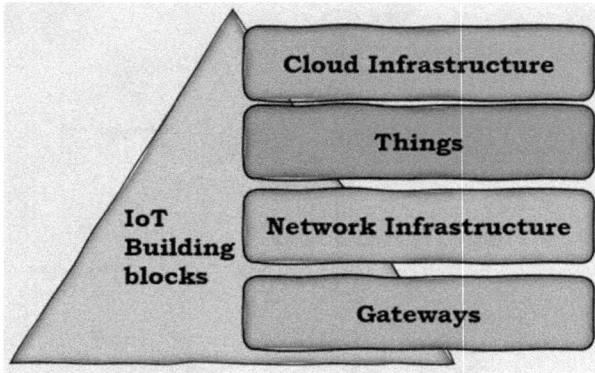

Figure 7.3 Basic components of IoT.

temperature and humidity. One of the IoT sensors involved in smart buildings is used to regulate the lighting by turning on and off as required.

7.2.6 Building automation systems (BAS)

The AI-based smart building system is shown in Figure 7.4. The main purpose of BAS is to control the multiple functions and improve the resources. It also helps to add additional devices and also controls the robust communication. The time taken to install each sensor directly relates to the entire cost [24]. BAS is responsible for each equipment performance and manages the building without any issues.

7.2.7 IoT-based indoor localization

The algorithm of IoT-based indoor localization decides that the position of things should be hybrid. A problem-solving algorithm that is a hybrid algorithm combines two or more different algorithms with each other. IoT applications are primarily responsible for sensor-based monitoring and deployment, and understanding the characteristics of the interior environment is the key to increasing efficiency [25].

7.2.8 Lighting system

Different categories of IoT systems enabled by AI are shown in Figure 7.5. It is very important to implement an IoT-based lighting system in smart buildings in order to control the overall lighting system in smart buildings. In future, visible light communication (VLC)-based solutions will be a great framework in the sector of IoT. The system also depends on infrared transmission where the indoor and outdoor is formed by a free-space optical communication system. VLC is generally implemented in the visible light from LEDs.

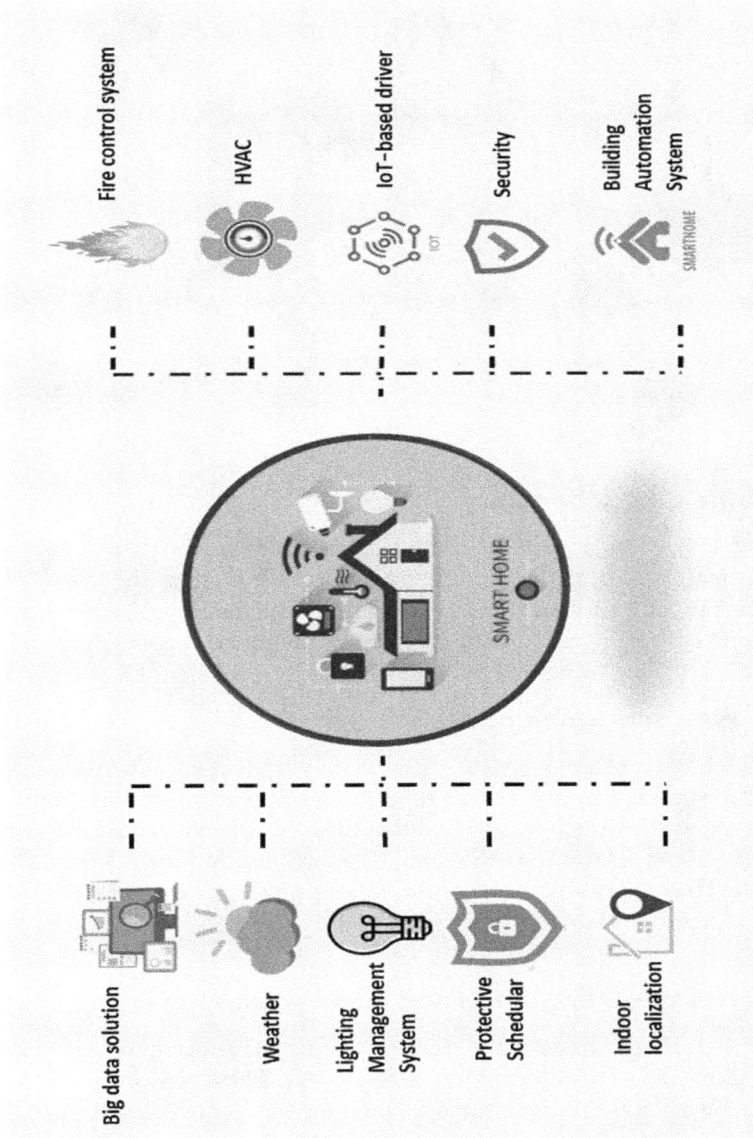

Figure 7.4 Artificial intelligence-based systems for IoT.

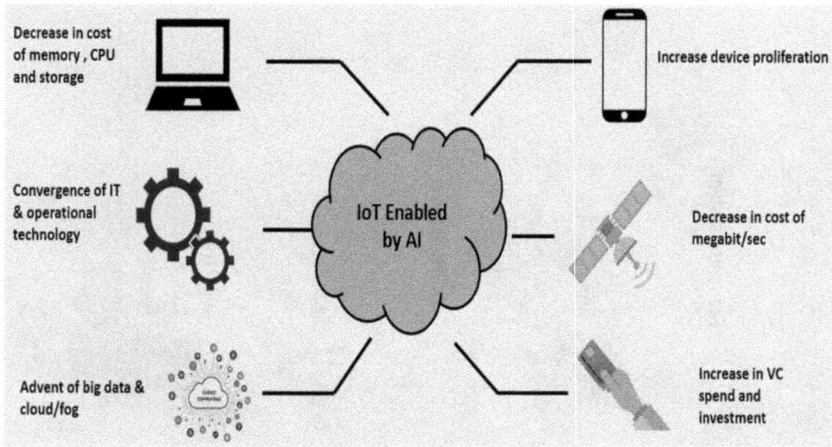

Figure 7.5 Different categories of IoT systems enabled by AI.

7.2.9 Shielding schedulers

Shielding schedulers are nothing but a smart plug which detects and controls the major load; if the amount of load exceeds, then the smart plug will automatically turn off [26]. The lighting and building's management systems have the centralized control of smart plug load schedulers.

7.2.10 Fire management system

The IoT sensor is responsible for the accurate delivery of information and to find the specific location where the incident happens. This fire management system is controlled by app, cloud and wirelessly connected devices. It is essential to implement the fire sensor in smart buildings to enhance safety measures. These sensors can automatically sense smoke and fire, and produce an alarm sound to alert the users.

7.2.11 Heating, ventilation and air conditioning

Heating, ventilation and air conditioning (HVAC) sensors will provide information about the environment and surroundings. These sensors are directly associated with automation system of buildings [27]. HVAC tools provide comfort to the user by detecting the temperature, airflow and humidity in various zone of buildings. The improvisation in the safety measures is an important thing to reduce disasters; by that way, the HVAC helps to reduce the incidents by pre-detecting the occurrence.

7.2.12 Energy efficiency toward smart buildings

Improvising the energy efficiency in buildings is a combining task that comprises the entire lifecycle of the building. The important stages to optimize the energy efficiency are as follows:

- Design
- Construction
- Operation
- Maintenance
- Demolition

To adapt the energy efficiency management system to the condition of building, it is very important to frequently reengineer the indexes that are used to measure the efficiency of energy. There are static and dynamic cases, which will influence the energy consumption of buildings. Depending upon the design, the static model will vary for each building. The researchers are mainly focusing toward solving the issues in the dynamic condition in the buildings.

7.2.13 Monitoring

The information is collected from the heterogeneous sources and diagnosed in advance to proposing concrete task to reduce the consumption of energy with respect to significant context of buildings. Based on the functionality of buildings, the energy usage will be different. It is essential to carry off a beginning characterization of the major contributors to increase their energy usage. For example, in residential buildings, the major energy consumption is due to the indoor system, and coming to industrial buildings, the energy consumption will increase due to the industrial machines [28]. The important parameters that should be monitored and diagnosed before installing most appropriate energy building management systems are as follows:

- Electrical devices should always be connected to the electrical network.
- Resident' behaviour.
- Geographical conditions.
- Information based on energy generated.
- Information based on total energy consumption.

7.2.14 Information management

The main aim of the intelligent management system is to provide actual adaptation measures for both devices and users; it also works to provide building comfort and requirement of energy efficiency [29]. Hence, energy saving totally depends on the quality of services in building and requirement of energy resource of building.

7.2.15 Automation system

The sensors used in the corridors and rooms in buildings will send data as input and these data will control the subsystems like HVAC system and security. This overall system is known as automation system. This system is also used to save energy and to estimate the environmental parameter and location of users. Therefore, implementation of an automation system in smart buildings is the most essential factor.

7.2.16 Feedback

Feedback based on energy consumption is important for energy saving. The smart metering will provide a real-time feedback on household energy utilization. From this smart metering, we came to know that the energy monitoring system can help to reduce 5% to 15% of energy consumption. A set of subsystems helps to give information about energy consumption in an essential way. Few subsystems are boilers, electrical panels, lighting and so on.

7.2.17 User participation

A million of users who had settled up with their traditional way of energy bills were not aware about the building of energy consumption till now. The users must need to know about the opportunities in installing smart building. The improvement of any intelligent system is in the hand of different users; the smart building also achieves the next level only through the interaction, data and feedback from the users.

7.3 SMART BUILDING ARCHITECTURE

Nowadays, automation in buildings is considered as the important aspect for the whole world. These smart buildings are necessary to save time for humans. IoT helps people to facilitate the new types of services in their everyday life [30]. With the help of modern technologies such as Bigdata analysis, Data science and Cloud computing, a smart building can be built. BAS is the main part to make a smart building. In the smart buildings, a computer monitors the whole building with the use of wireless sensors and without the help of humans. This automation system controls all the electrical equipment without using any manpower. Smart buildings help the society by increasing energy optimisation techniques.

Protective lock system, smart energy management system, indoor localisation system, fire control system, HVAC, movement monitoring system, weather system, cloud infrastructure and network infrastructure are AI-based systems which are used in smart buildings. All these systems are integrated together to maintain smart buildings with the use of ML algorithms [31]. The sensors and microchips in the buildings are useful to collect and store the data about the services and functions in smart buildings. IoT sensors, analytical software, user interface and connectivity are the four basic components in smart buildings. Smart buildings are more advanced than Command and Control mechanisms. For instance, Building Management System helps (BMS) to turn ON HVAC when the level of carbondioxide is high inside the smart building and turns OFF HVAC automatically when the level of CO_2 is reduced lower than the limit. BMS also reduces the usage of high energy and decreases the energy bills for the smart building users. They keep on monitoring the whole control of the building to reduce the risk of life of the living creature inside the building [32].

The main objective of these smart buildings is to free up the stress of humans and to consume energy by including new and advanced features in the buildings. The rapid growth in the population will lead to the increase in occupancy of the buildings. At that time, smart buildings will play a major role in the world.

7.3.1 Building Energy Management System (BEMS)

The BEMS is a suitable method to monitor the needs of the energy in smart buildings. The BEMS architecture is shown in Figure 7.6. This system differs from other systems by communication characteristics. This system receives the functions of all other systems used in the building which help to process all other system works in smart buildings. There are three main components used in effective energy management strategy. They are risk management, efficiency, environmental sustainability. First of all, the building energy management system (BEMS) identifies the sources of energy consumption. In this breaking down, the high energy consumption helps smart buildings to identify where the energy is consumed at most. This is useful for both the common and to the government, as the energy bills are reduced, which plays a role in the energy optimisation.

Energy conservation and demand side management (DSM) are two sides of energy management [33]. DSM is the energy managing activity, which describes to reduce the use of high energy during the peak hours of energy usage. The importance of these systems is to reduce the high energy usage because the cost to produce electricity is high. Automation in technologies increases day by day, as the utilisation of energy increases. Building management system firms may increase the calibre and efficacy of solutions to the escalating energy demand issue by adopting automation systems in smart buildings with current infrastructure.

Maintenance plays a vital role in BEMS. Without continuous maintenance like the replacement of batteries for sensors, checking for proper

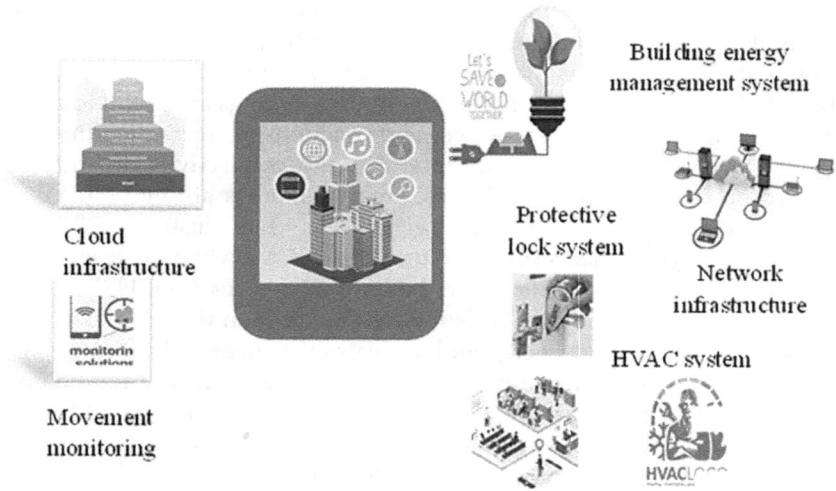

Figure 7.6 Building energy management system architecture.

connections, updating the software for security and speed purposes, checking the operation of sensors and computers used in smart buildings [34]. The three main ways to optimise energy inside smart buildings are energy conservation, energy recovery and energy substitution. Energy conservation is the process of reducing the energy waste, through more rational use. Energy substitution refers to the process of replacing the conventional source of energy with a nonconventional source for smart buildings [35]. The BEMS must have a control system where all the systems used in smart buildings can be controlled with a single application. This system software must take into account dimensions of the smart building and the interconnections inside the building.

7.3.2 Protective lock system

Protective lock system in smart buildings consists of fingerprint lock, voice and video intercommunication system, code-based access system, swipe card access system, and surveillance through CCTVs (Closed-Circuit Television). This protective system helps to restrict the entry and exit of unknown people to the buildings. Cameras must surround the whole building without leaving any blind spot. Because blind spots are the main cause for the occurrence of physical attacks, biometrics and radiofrequency identification (RFID) systems are the fastest growing sectors in automatic identification sectors. People must be aware about linking of physical and digital security as it cannot be easily foolished [36]. As of now, most of the industrial data theft is done only by physical thefts. By the use of digital security systems, the physical thefts can be controlled and the whole building can be monitored easily. By including the digital security systems, the cost can be reduced, as the use of securities are reduced.

7.3.3 Fire safety system

Fire accidents in smart buildings can be controlled by the use of sprinkler systems. Sprinkler systems help people to escape from fire and it supresses the fire by spraying a large amount of water. The sensor in the fire alarm system will detect the fire and passes the message to the actuators with the help of a transmitter. The actuators receive the signal and act as a switch to turn on the sprinkler system. Using the sensors, the sprinkler system points over the fire and will stop automatically after the room reaches the sustained temperature. A sprinkler will emit nearly 50 litres of water per minute; around 75% of fire will be evaporated, as fire brigade uses around 300 gallons per minute [37]. Nowadays, people have the misconception that the sprinkler looks ugly in the neat resident area and the water that is sprayed by the sprinkler will ruin the whole place.

7.3.4 HVAC

Heating and air conditioning are the major cost increasing factors in the construction of buildings. With the help of the HVAC system in smart buildings, the cost and energy can be reduced as compared to the cost and energy consumed in normal residential areas. With the help of IoT, the HVAC system turns ON and OFF automatically by the sensors used to measure the temperature inside the building. By this IoT-enabled system, the energy can be optimised. When the measured heat exceeds the thresholds, IoT sends alert to the concerned systems.

7.3.5 Movement monitoring system

To monitor the movement inside smart buildings, various types of sensors can be used. Some of them are thermal sensors, motion or occupancy sensors. Both of these sensors capture the image of the person, animal or an object, which is physical movement inside the room. In the olden days, motion sensors were used in the industries to capture the image of the intruders who were included in illegal activities inside the industry. Nowadays, these motion sensors are used by common man who lives in the society. Motion sensors work by detecting the infrared radiation or by the reflection of ultrasonic waves that is emitted by the device [38]. These sensors detect the heat emitted by the people and start capturing the movement. These sensors sense and transform the information into an electric signal, which is transmitted to the master device.

Fire safety system

24/7 monitor system

Updated security system

Energy optimisation

Improved Insulation

Rainwater utilisation system

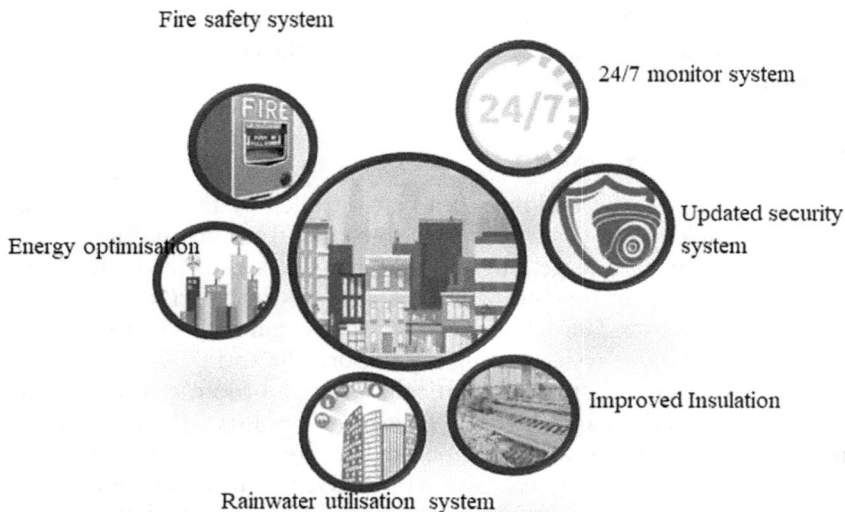

Figure 7.7 Features of smart building.

7.3.6 Cloud infrastructure

Cloud-based BMS helps to allow the person to access the whole building process from any place of earth with an internet-enabled device. The advanced features of smart building are shown in Figure 7.7. With the cloud infrastructure, the data collected from all the systems can be uploaded to the internet. By this system, the security of the data can be increased. An intruder cannot access or is not able to make any change on the smart building without the access id and password. The data can be taken easily on the critical purposes within a small period of time. This system helps to increase the energy efficiency, reduce manpower, reduce operating costs and increase security [39].

7.4 MONITORING TECHNIQUES FOR SMART BUILDINGS

With the help of IoT devices like smartphone, cameras, sensors, etc., it is much easier to monitor and collect the required data and information. The data and information can be collected very easily by using IOT devices. Energy consumption in a smart building can be improved by rectifying the occupancy monitoring problems [40]. Researchers have suggested some solutions for the occupancy-related problems. We have focused on certain challenges and criteria to identify the problem. They areas follows:

1) To achieve occupancy monitoring technique from the existing infrastructure of the building.
2) To achieve occupancy process in a miniature way.
3) To execute occupancy monitoring process without using any external applications or new technologies.
4) To achieve a way for the existing infrastructure of the smart buildings without any external applications.
5) To develop occupancy monitoring technique by using data fusion process.

In occupancy monitoring process, there will be several variations and problems. There are several categories, but they are interrelated to each other to provide a better solution for improvement of smart buildings. There are four categories in occupancy monitoring process, which are described below.

7.4.1 Occupancy detection

This method is used to detect whether the free space in a particular area is occupied or not and it detects the availability of free space within the time limit. This technique does not reveal how much free space is occupied or how many people have been seated; it gives the desired data and its availability by using binary numbers [41]. The spaces occupancy detection for the public places like cafeteria, theatre, multipurpose hall is much difficult than for houses. So, these places can be monitored by using advanced terminologies like cameras, which is one of the examples.

7.4.2 Occupancy counting

The main aim of this technique is to count how many people are there in a building at a certain period of time. It can be divided into two types:

1) Counting by whole
2) Counting at a specific place or zone.

7.4.2.1 Counting by whole

In this type of counting process, the people at the whole building will be counted. It helps to determine how many people are there in a certain place or a building in a specific time.

7.4.2.2 Counting at a specific place or zone

This type of counting takes place at different levels. This process analyses the building first and it divides into several categories like Wi-Fi enabled zone, drinking area, kitchen area, dining area and so on.

7.4.3 Occupancy tracking

This process detects, locates, counts, observes and even tracks the status of the people. This process is the integration of the above two processes. This process is accurate and is much useful in smart buildings. By framing algorithms, the solutions for the problems can be obtained more easily.

7.4.4 Occupancy event recognition

This process detects the movement of an individual at a certain location. This process has the ability to detect and monitor the behavioural activities of an individual. These data are collective and informative for the future use. This process is able to detect a single individual or even a group of people more accurately within a shortest period by observing their activities and actions.

When looking into the problems of the smart buildings, researchers have proposed some techniques. They are:

1) By using existing Wi-Fi setup and without adding any external networks or applications.
2) By using external software, applications, some of the advanced terminological devices, etc...
3) By using sensors, cameras, etc.

In recent times, people construct the smart buildings based on the HVAC system to provide a comfort zone for the occupants. This system is designed to achieve a more spacious and well-organised environment for the occupants. The HVAC systems are classified into four categories, namely:

1) *Heating and cooling split systems* – This system has two separate units for heating and cooling purposes.
2) *Hybrid split type* – This system is a hybrid heating type, and it also helps to reduce the energy costs in a building.
3) *Duct free* – This type is more suitable for hotels, party halls, etc. This type functions by inserting duct free units in certain areas of a place and it provides heating or cooling when they are needed.
4) *Packaged heating and air conditioning system* – This type consists of all the compressors, condensers and evaluators in a single space.

The Wi-Fi based monitoring system for smart buildings is shown in Figure 7.8. These systems are designed based on the data and information collected by the various sensors and cameras. So, for designing these sensors, cameras and certain applications and hardware designs and setup, maintenance and integration of software and hardware requires equivalent costs. In addition to hardware costs, the maintenance problems and their costs are quite larger

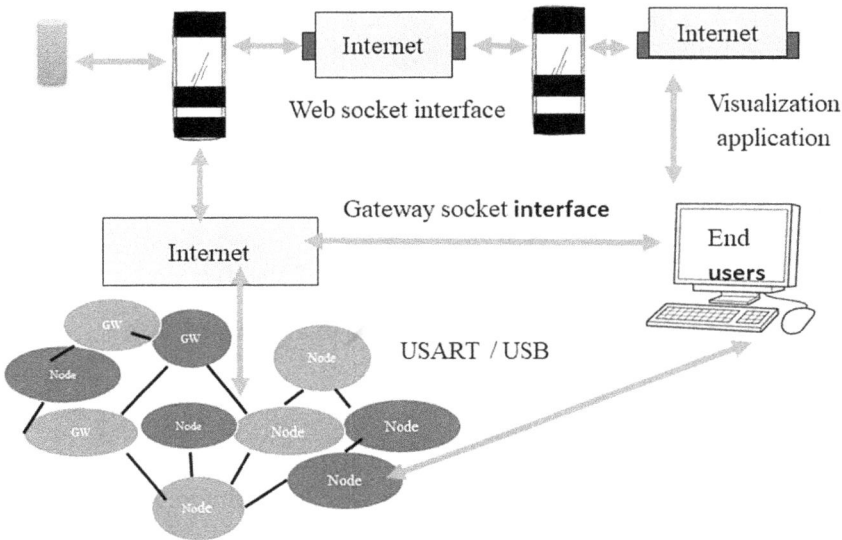

Figure 7.8 Wi-Fi based monitoring system for smart buildings.

than the others. So, most of the people avoid smart buildings because of these reasons. So, to avoid these kinds of issues, the researchers have suggested to develop smart buildings using the existing Wi-Fi network services without introducing any external applications or software [42]. At present, many people are enabling Wi-Fi networks for many purposes. So, this enables to adapt the individual for the smart buildings. However, an individual will have smart devices like mobile phones, smart watches and so on with him. This helps to monitor and detect the comfort zone in our home.

There are several ways in this smart building architecture by which an individual even can turn on and turn off his microwave oven, air conditioners by a single click in his mobile phone through an application. Nowadays, fingerprint-based techniques have been launched like fingerprint sensor automated doors. These doors do not need any lock; if we sense our fingerprint in the doors, it gets locked and opened automatically. These fingerprint-based sensors are very much useful. There are several processes carried out by Received signal strength indicator (RSSI). This process needs a packet analyser to be installed to monitor the incoming and outgoing packets, which are stored in computer to extract MAC address and the received signal strength indicator values.

Recently, some researchers have worked on Authorization, Authentication and Accounting (AAA) techniques. They have used Wi-Fi network and a smartphone. A separate application will be designed for this process to authorise, authenticate and account the data and information, which were obtained. With the use of Wi-Fi network and application, MAC address can be identified more accurately. So, these processes can be integrated with the HVAC system. This provides more accuracy and effectiveness. The researchers have performed many experiments on this technique and have obtained 86% of good results. The system has obtained a good accurate level. When there are any inaccuracies, the Wi-Fi network will get automatically turned off. The researchers have reported that they have consumed 18% of energy by using this technique.

7.4.5 Sensor network-based occupancy monitoring technique

There are several sensors, which were used in the smart buildings like humidity sensors, temperature sensors, pressure sensors, CO_2 sensors, ultrasonic sensors and so on. These sensors are much useful in the advancement of smart building architecture. There are three important machine learning techniques that are useful to analyse and process the data from a sensor network. They are much useful to collect the information about the carbon dioxide level in the air, lighting, temperature, humidity level.

7.4.6 CO_2 sensors

An individual will release a certain concentration of carbon dioxide in the air when he inhales or exhales the air. These are measures using parts per millions (ppm) [43]. These sensors are easier to install and can be reused many times. But these sensors need certain amount of time to obtain accurate results. These sensors should be chosen based on the range, cost, drift, longevity and sensitivity.

These CO_2 sensors are classified into three types, namely:

1) Nondispersive infrared sensors
2) Electrochemical sensors
3) Metal oxide semiconductor sensors

7.4.6.1 Nondispersive infrared sensors

These sensors measure the amount of CO_2 in the air by using specific wavelength of light. Each and every element absorbs a specific amount of light source, and it breaks into atoms and molecules. So, we can obtain the results of how much amount of light is observed by each particle. When the air enters the sensor, the wavelength of light is passed through the carbon dioxide particles. The other end is used to detect how much amount of light is observed by a particle. If there is more concentration level of carbon dioxide, there will be more light source observed.

7.4.6.2 Electrochemical sensors

These sensors are used to measure the electric current or conductivity level to obtain the carbon dioxide level present in the air. When the carbon dioxide enters the sensor, the chemical process takes place between the carbon dioxide and electrical charge. When the electric charge is consumed at higher rates, the carbon dioxide level is higher.

7.4.6.3 Metal oxide semiconductor sensors

These sensors use resistive element for the free flowing of electric current through the metal strip. When the air enters through the metal strip, it gets in contact with the electric current due to chemical reaction between the metal strip and electric current. There will be change in the chemical composition by oxidation or reduction process. This detects how much amount of gas is absorbed by the particle to obtain the carbon dioxide level.

Ultrasonic sensors use radar signal to transmit from transmitter to the ultrasonic chirps, which are reflected from the humans. The microphone receives the signals, and they detect and monitor the data. The time taken to receive signals from the transmitter is the main criteria in the ultrasonic sensor.

7.4.7 Humidity sensors

Humidity sensors detect the quantity of water vapour in the atmosphere. As too much moisture can corrode some machineries [44], these sensors are used in museums, church, hospitals, hotels and manufacturing plants.

There are three types of humidity sensors, namely,

1) *Capacitive* – These sensors have porous dielectric substance at the centre. There will be two electrodes, which act as a protective layer for the dielectric substance. Water vapour is the main component to check the humidity level. When the water vapour flows through the electrodes due to the chemical reaction, it creates a voltage change.
2) *Resistive* – These sensors are less sensitive than capacitive sensors. These sensors measure humidity level due to electrical change. These sensors use ions to measure the resistance in the electrodes.
3) *Thermal* – There will be two thermal sensors, which conduct electricity based on the humidity level in the air. One sensor is coated in dry nitrogen and the other sensor measures the air. The difference in values of the sensors is used to obtain the accurate humidity level in the air.

7.4.8 Temperature sensor

These sensors are used to measure the heat and detect and analyse the changes in the temperature. These sensors help for business environment to provide automatic HVAC control to maintain the correct level.

There are four types of temperature sensors; they are as follows:

1) *Semiconductor-based sensors* – These sensors use diodes, which are designed with a temperature-sensitive component. The difference in voltage levels is used to record the changes in temperature levels.
2) *Thermocouple* – It is made up of two wires composed of different metals that are placed at various locations. Temperature variations are caused by changes in two places and the voltage distribution.
3) *Resistance temperature detector* – These sensors measure the temperature at the great accuracy level. These sensors are more expensive than the other type of sensors. A ceramic substance is wrapped with a wire. The changes in resistance of the wire are used to measure the temperature.
4) *Negative temperature coefficient thermistor* – When the temperature increases, the resistance drops quickly. These sensors are used to measure the resistance more accurately.

7.4.9 Microphone

The noise produced by the people in a specific area or place can be captured and analysed using microphone. The position of the noise source can be detected using an array of microphones in a particular area. Every device will have an inbuilt microphone feature in them. Speech recognition and noise segregation are some of the examples.

The above mentioned are some of the methods and processes in sensor monitoring technique. There are multiple sensors in the market. This topic explains about CO_2 sensors, ultrasonic sensors, humidity sensors, temperature sensors and microphone.

7.4.10 Data fusion for occupancy tracking

Indoor environment can be improved by using data fusion techniques, which can simultaneously detect the data and information obtained from various types of sensors [45]. There are some terminologies in data fusion techniques; they are as follows:

1) Data association – Data fusion schemes are used to correlate the information and data.
2) State estimation – In this process, the information obtained from the various sensors at different levels are used to reach the high state estimation accuracy level.
3) Classification – This process consists of clustering data, which are divided into different categories by comparing the characteristics and behavioural stages.

4) Prediction – This process deals with simple processes. It is mainly used to obtain and predict outputs for the obtained information sources.
5) Unsupervised machine learning – This process consists of unsupervised machine learning algorithms. These algorithms are grouped in this process.
6) Dimensionality reduction – This process involves reduction in dimensionality of the data and information, which were obtained from multiple sensors. This also decreases computational time and also yield high- dimensional data.
7) Statistical interference – This process is used to outline some more additional features in addition to the existing one.
8) Visualization – This process is used to provide the statistical data of energy consumption in smart buildings to the consumers. Different varieties of data sources are fused to obtain a powerful visualisation tool.

7.4.11 Camera-based monitoring

Camera-based people counting is classified into three types; they are as follows:

1) Extracting features based on counting the number of people
2) Extracting features by tracking the moving region
3) Extracting features and estimating the number of people directly

The first method is very difficult, as we cannot count every person directly because it much challenging. The second method helps to overcome the barricade problems, which were common in the first method. This method also has several complexities like variations in motion paths by different people or a moving body and there will be similarities of patterns between several people. The next method is very useful in counting the number of people [46]. This method does not provide information and data about where the people lived, located and survived and so on. The chances of fusion of data in regression method aremore difficult. This method provides only the exact count. The body parts which are mentioned for counting are the head, face, shoulder, upper body and skeleton. The algorithms are obtained only when the object is at motion. For example, when a correspondent in a school is conducting meeting for teachers for three hours, then, the data for three hours will not be accurate because the people or an object at rest can be detected accurately. This leads to less accuracy in the outcome.

The camera monitoring technique will only have simple algorithms for the recognition of the object in which the information and data are sent to the data server after these processes [47]. These are some monitoring techniques in smart buildings.

7.5 OPPORTUNITIES AND CHALLENGES

The market for smart buildings now offers the maximum prospects. One of the most crucial applications of IoT technology is smart buildings. However, there are several obstacles to IoT growth in smart buildings [26].

They are:

- Connectivity
- Security

7.5.1 Connectivity

One of the biggest challenges is to connect and communicate in different systems. Smart building is based on data collection of different devices and sensors in the building. So, there is a need to secure the data and connect with different devices. If installation and connection process is finished successfully, we can control the building anywhere and anytime. By using a smartphone, we will control all devices in the smart building.

7.5.2 Security

One of the biggest challenge is the need to secure all data from hackers. We need to ensure authenticity of data, integrity of data and also privacy. As such, it is vital that every IoT device connected to a network is carefully monitored and patched effectively as soon as an issue occurs. It effectively monitors and maintains connected IoT devices, ensuring any potential risk alert is patched as soon as possible. Put measures in place to stop unauthorized users from accessing devices, data or your network, without encryption key. If a security risk is identified, have a pre-prepared 'multi-layer' security defence strategy in place to before threat.

Some IoT security challenges in a smart building are

- Poor testing
- Weak password
- Lack of visibility

7.5.3 Poor testing

Because most IoT developers do not prioritise security, they find it difficult to identify weaknesses in IoT systems.

7.5.4 Weak password

Cybercriminals have easy access to IoT devices because they frequently come with default passwords that many users forget to update. The remaining users generate easily guessable passwords.

7.5.5 Lack of visibility

It is hard to maintain a precise inventory of what needs to be safeguarded and monitored due to users utilising IoT devices without IT departments' awareness.

7.6 CONCLUSION

IoT techniques are being sought-after by researchers in the field of smart buildings for controlling, assessing and enhancing energy efficiency. The most significant aspects of smart buildings are covered in this chapter, with a focus on what is now expected of them and why IoT is crucial for making buildings energy efficient. There are many benefits to using IoT technology in smart buildings, but there are challenges as well. An overview of IoT technologies, architecture and smart building monitoring has been provided in this chapter. Despite recent technology developments in IoT making it feasible to execute the idea of smart buildings, there are still a number of issues with energy efficiency in smart buildings. It will be a huge motivator for advancements in both the academic and commercial areas of smart building research if we are able to quickly identify additional solutions to many problems.

REFERENCES

1. Akkaya, K., Guvenc, I., Aygun, R., Pala, N., & Kadri, A. (2015, March). IoT-based occupancy monitoring techniques for energy-efficient smart buildings. In *2015 IEEE wireless communications and networking conference workshops (WCNCW)* (pp. 58–63). IEEE.
2. Kumar, A., Sharma, S., Goyal, N., Singh, A., Cheng, X., & Singh, P. (2021). Secure and energy-efficient smart building architecture with emerging technology IoT. *Computer Communications, 176*, 207–217.
3. King, J., & Perry, C. (2017). *Smart buildings: Using smart technology to save energy in existing buildings*. Washington, DC: Amercian Council for an Energy-Efficient Economy.
4. Farzaneh, H., Malehmirchegini, L., Bejan, A., Afolabi, T., Mulumba, A., & Daka, P. P. (2021). Artificial intelligence evolution in smart buildings for energy efficiency. *Applied Sciences, 11*(2), 763.
5. Leninpugalhanthi, P., Janani, R., Nidheesh, S., Mamtha, R. V., Keerthana, I., & Kumar, R. S. (2019). Power theft identification system using IoT. In *2019 5th international conference on advanced computing & communication systems (ICACCS)* (pp. 825–830). IEEE.
6. Aljafari, B., Ramu, S. K., Devarajan, G., & Vairavasundaram, I. (2022). Integration of photovoltaic-based transformerless high step-up dual-output–dual-input converter with low power losses for energy storage applications. *Energies, 15*(15), 5559.

7. Dhanyaa, N., Kumar, R. S., Adhithya, V., Leninpugalhanthi, P., Kishoreadhithyaa, B., Ishwarya, S., & Kaviyanjali, V. (2022, March). A review of plant disease detection techniques using artificial intelligence. In *2022 8th international conference on advanced computing and communication systems (ICACCS)* (Vol. 1, pp. 512–515). IEEE.

8. Leninpugalhanthi, P., Kumar, R. S., Ramya, G., Prithika, P., Nandhakishore, R., Narendaren, K., & Ishwarya, S. (2022, March). Lithium-ion battery life estimation using machine learning algorithm. In *2022 8th international conference on advanced computing and communication systems (ICACCS)* (Vol. 1, pp. 573–576). IEEE.

9. Ramu, S. K., Irudayaraj, G. C. R., & Elango, R. (2021). An IoT-based smart monitoring scheme for solar PV applications. *Electrical and Electronic Devices, Circuits, and Materials: Technological Challenges and Solutions*, 1, 111–119.

10. Kumar, R. S., Leninpugalhanthi, P., Rathika, S., Rithika, G., & Sandhya, S. (2021, March). Implementation of IoT based smart assistance gloves for disabled people. In *2021 7th international conference on advanced computing and communication systems (ICACCS)* (Vol. 1, pp. 1160–1164). IEEE.

11. Saravanan, S., Kumar, B. R. R., Kumar, S. R., Sarudharshini, R., & Shankar, S. (2022). Methodology for low-cost self reconfiguration process. In *2022 international conference on inventive computation technologies (ICICT)* (pp. 570–574). IEEE.

12. Singh, P., Arora, K., & Rathore, U. C. 2022. Control strategies for improvement of power quality in grid connected variable speed WECS with DFIG–An overview. *Journal of Physics: Conference Series*, 1, 32–41.

13. Aliero, M. S., Asif, M., Ghani, I., Pasha, M. F., & Jeong, S. R. (2022). Systematic review analysis on smart building: Challenges and opportunities. *Sustainability*, 14(5), 3009.

14. Barone, G., Zacharopoulos, A., Buonomano, A., Forzano, C., Giuzio, G. F., Mondol, J., ... & Smyth, M. (2022). Concentrating photovoltaic glazing (CoPVG) system: Modelling and simulation of smart building façade. *Energy*, 238, 121597.

15. Moreno, M. V., Úbeda, B., Skarmeta, A. F., & Zamora, M. A. (2014). How can we tackle energy efficiency in IoT based smart buildings? *Sensors*, 14(6), 9582–9614.

16. Metallidou, C. K., Psannis, K. E., & Egyptiadou, E. A. (2020). Energy efficiency in smart buildings: IoT approaches. *IEEE Access*, 8, 63679–63699.

17. Majdi, A., Dwijendra, N. K. A., Muda, I., Chetthamrongchai, P., Sivaraman, R., & Hammid, A. T. (2022). A smart building with integrated energy management: Steps toward the creation of a smart city. *Sustainable Energy Technologies and Assessments*, 53, 102663.

18. Shah, S. F. A., Iqbal, M., Aziz, Z., Rana, T. A., Khalid, A., Cheah, Y. N., & Arif, M. (2022). The role of machine learning and the internet of things in smart buildings for energy efficiency. *Applied Sciences*, 12(15), 7882.

19. Lin, Q., Chen, Y. C., Chen, F., DeGanyar, T., & Yin, H. (2022). Design and experiments of a thermoelectric-powered wireless sensor network platform for smart building envelope. *Applied Energy*, 305, 117791.

20. Dong, B., Prakash, V., Feng, F., & O'Neill, Z. (2019). A review of smart building sensing system for better indoor environment control. *Energy and Buildings*, 199, 29–46.

21. Kumar, R. S., Raj, I. G. C., Saravanan, S., Leninpugalhanthi, P., & Pandiyan, P. (2021). Impact of power quality issues in residential systems. In *Power quality in modern power systems* (pp. 163–191). USA: Academic Press.

22. Shah, A. S., Nasir, H., Fayaz, M., Lajis, A., & Shah, A. (2019). A review on energy consumption optimization techniques in IoT based smart building environments. *Information*, 10(3), 108.

23. Ghayvat, H., Mukhopadhyay, S., Gui, X., & Suryadevara, N. (2015). WSN- and IOT-based smart homes and their extension to smart buildings. *Sensors*, 15(5), 10350–10379.

24. Park, H., & Rhee, S. B. (2018). IoT-based smart building environment service for occupants' thermal comfort. *Journal of Sensors*, 1, 233–245.

25. Panchalingam, R., & Chan, K. C. (2021). A state-of-the-art review on artificial intelligence for smart buildings. *Intelligent Buildings International*, 13(4), 203–226.

26. Bagheri, A., Genikomsakis, K. N., Koutra, S., Sakellariou, V., & Ioakimidis, C. S. (2021). Use of AI algorithms in different building typologies for energy efficiency towards smart buildings. *Buildings*, 11(12), 613.

27. Saran, S., Gurjar, M., Baronia, A., Sivapurapu, V., Ghosh, P. S., Raju, G. M., & Maurya, I. (2020). Heating, ventilation and air conditioning (HVAC) in intensive care unit. *Critical Care*, 24(1), 1–11.

28. Carli, R., Cavone, G., Ben Othman, S., & Dotoli, M. (2020). IoT based architecture for model predictive control of HVAC systems in smart buildings. *Sensors*, 20(3), 781.

29. Minoli, D., Sohraby, K., & Occhiogrosso, B. (2017). IoT considerations, requirements, and architectures for smart buildings—Energy optimization and next-generation building management systems. *IEEE Internet of Things Journal*, 4(1), 269–283.

30. Abbasi, A. Z., Islam, N., & Shaikh, Z. A. (2014). A review of wireless sensors and networks' applications in agriculture. *Computer Standards & Interfaces*, 36(2), 263–270.

31. Le, D. N., Le Tuan, L., & Tuan, M. N. D. (2019). Smart-building management system: An Internet-of-Things (IoT) application business model in Vietnam. *Technological Forecasting and Social Change*, 141, 22–35.

32. Chen, H., Chou, P., Duri, S., Lei, H., & Reason, J. (2009, October). The design and implementation of a smart building control system. In *2009 IEEE international conference on e-business engineering* (Buckman, A. H., Mayfield, M., & Beck, S. B. (2014). What is a smart building? *Smart and sustainable built environment*).

33. Buckman, A. H., Mayfield, M., & Beck, S. B. (2014). What is a smart building? *Smart and Sustainable Built Environment*, 1, 356–364.

34. Verma, A., Prakash, S., Srivastava, V., Kumar, A., & Mukhopadhyay, S. C. (2019). Sensing, controlling, and IoT infrastructure in smart building: A review. *IEEE Sensors Journal*, 19(20), 9036–9046.

35. Ramu, S. K., Irudayaraj, G. C. R., Subramani, S., & Subramaniam, U. (2020). Broken rotor bar fault detection using Hilbert transform and neural networks applied to direct torque control of induction motor drive. *IET Power Electronics*, 13(15), 3328–3338.

36. Rahman, A., Nasir, M. K., Rahman, Z., Mosavi, A., Shahab, S., & Minaei-Bidgoli, B. (2020). Distblockbuilding: A distributed blockchain-based sdn-iot network for smart building management. *IEEE Access*, 8, 140008–140018.

37. Edirisinghe, R., & Woo, J. (2020). BIM-based performance monitoring for smart building management. *Facilities, 1*, 366–371.
38. Irwin, D., Barker, S., Mishra, A., Shenoy, P., Wu, A., & Albrecht, J. (2011, November). Exploiting home automation protocols for load monitoring in smart buildings. In *Proceedings of the third ACM workshop on embedded sensing systems for energy-efficiency in buildings* (pp. 7–12). IEEE proceedings.
39. Vattapparamban, E., Çiftler, B. S., Güvenç, I., Akkaya, K., & Kadri, A. (2016, May). Indoor occupancy tracking in smart buildings using passive sniffing of probe requests. In *2016 IEEE international conference on communications workshops (ICC)* (pp. 38–44). IEEE.
40. Moreno, M., Dufour, L., Skarmeta, A. F., Jara, A. J., Genoud, D., Ladevie, B., & Bezian, J. J. (2016). Big data: The key to energy efficiency in smart buildings. *Soft Computing, 20*(5), 1749–1762.
41. Aliero, M. S., Qureshi, K. N., Pasha, M. F., Ghani, I., & Yauri, R. A. (2021). Systematic mapping study on energy optimization solutions in smart building structure: Opportunities and challenges. *Wireless Personal Communications, 119*(3), 2017–2053.
42. Aguilar, J., Garces-Jimenez, A., R-Moreno, M. D., & García, R. (2021). A systematic literature review on the use of artificial intelligence in energy self-management in smart buildings. *Renewable and Sustainable Energy Reviews, 151*, 111530.
43. Moreno, M. V., Zamora, M. A., & Skarmeta, A. F. (2014). User-centric smart buildings for energy sustainable smart cities. *Transactions on Emerging Telecommunications Technologies, 25*(1), 41–55.
44. Jia, M., Komeily, A., Wang, Y., & Srinivasan, R. S. (2019). Adopting Internet of Things for the development of smart buildings: A review of enabling technologies and applications. *Automation in Construction, 101*, 111–126.
45. Benavente-Peces, C. (2019). On the energy efficiency in the next generation of smart buildings—Supporting technologies and techniques. *Energies, 12*(22), 4399.
46. Fotopoulou, E., Zafeiropoulos, A., Terroso-Sáenz, F., Şimşek, U., González-Vidal, A., Tsiolis, G., … & Skarmeta, A. (2017). Providing personalized energy management and awareness services for energy efficiency in smart buildings. *Sensors, 17*(9), 2054.
47. Casini, M. (2016). *Smart buildings: Advanced materials and nanotechnology to improve energy-efficiency and environmental performance*. USA: Woodhead Publishing.

Chapter 8

Soft computing techniques for renewable energy systems

Pradeep Singh Thakur

Rajiv Gandhi Government Engineering College, Kangra, India

Krishan Arora

Lovely Professional University, Phagwara, India

Umesh C. Rathore

Government Hydro Engineering College Bandla, Bilaspur, India

CONTENTS

8.1 INTRODUCTION

The complexity and size of electrical power system (PS) have increased. The narrow stability tolerance causes components of PS to perform significantly more closely to respective control and stability. Voltage unbalance can result from any faults or disturbances in the power transformers, power line, or machine used for generation [1]. Such modifications of the power

DOI: 10.1201/9781003407409-8

transmission parameter's working limit extend to the point where corrections equipment utilized in proposed control factors are unable to regulate them. Whenever a system experiences voltage stability issues, it causes the system's terminal voltage to drop, gradually rise, or surge when such system seems to be under heavy load [2]. In some countries like Germany, Sweden, Belgium, Japan and United States, there seems to be few incidents of voltage unbalance [3]. The increasing burden on the components of PS, which causes a change in voltage beyond their unified per unit cost, is the root cause of the electrical system's voltage regulation issue.

Building smarter and intelligent devices is the goal of the discipline of computer science known as SC (Soft Computing). The ability to extract the solution rather than just acquire it comes from intellect. As we move up the ranks, our capacity to handle complication, imprecise information, and pure thought as well as deep learning and liberty of movement all rise as shown in Figure 8.1. The ultimate goal is to create a software or system that functions similarly to how humans can, that is, a technique to artificially emulate human intelligence in devices. Another crucial element in machine learning is instinctive awareness or intelligence, which is constantly developed through contemplation. To integrate awareness in software is, in fact, a tremendous task and a very new occurrence. In order to attain resilience, controllability, and overall cheap cost, computational is a growing set of approaches that try to take advantage of acceptance for imperfection, ambiguity, and imprecise information. Computational techniques have proven useful in a variety of areas. Computational processes mimic awareness and

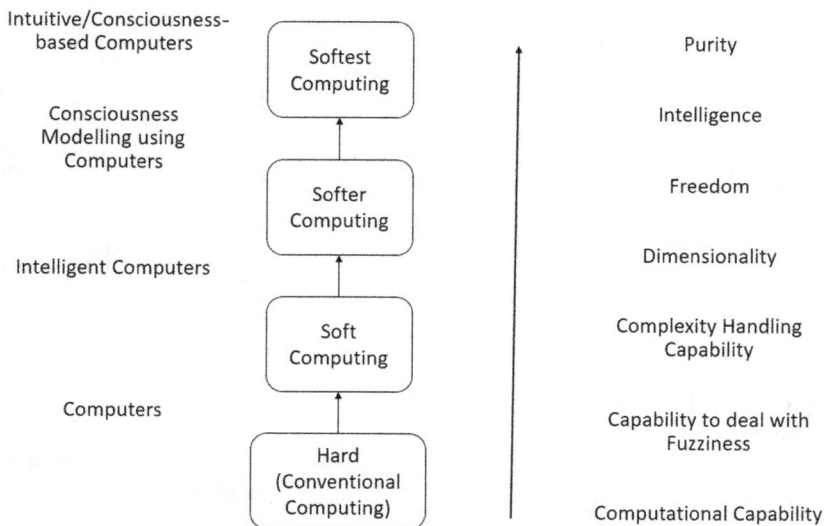

Figure 8.1 Development of soft computing.

comprehension in significant ways that differ from methodological approaches. Users could really take advantage of experience peoples, cognize into disciplines in which personal knowledge is lacking, and conduct input-to-output routing more quickly than intrinsically sequential analytical characterizations, thanks to multicore structures that illustrate biological mechanisms. Precision is reduced as an exchange, though. If a propensity for inaccuracies might be permitted, it really should be feasible to expand the technology range to include issues for which there are already widely obtainable quantitative and statistical interpretations. The anticipated reduction in computing burden and subsequent improvement in computing performances that enable better reliable systems serve as the impetus for this type of upgradation (Jang et al. 1997).

Lotfi A. Zadeh, the creator of fuzzy systems, coined the phrase "soft computing's" in 1994. A group of computational methods used in computer programming, intelligent systems, and several engineering fields that aim to explore, represent, and analyze highly complicated phenomena are collectively referred to as soft computing. This contrasts to traditional (hard) computation, which aims to maximize the latitude for ambiguity and inaccuracies in order to attain scalability, resilience, and smaller costs. A class of technique for solving the solution is known as "soft computing," which has similarities to genetic thinking and solution to problems (several times it is termed as perceptive computing) [4].

Zadeh characterized smart computing as the merger of fuzzy (FL) systems, neuro (NN) computing, adaptive and genetics computing, plus stochastic data processing together into single multi-professional system. Computation is the combination of approaches intended to analyze and facilitate answers to real-world issues that cannot be quantitatively modeled or are analytically extremely difficult to simulate. The goal of smart computational techniques is to replicate human judgment as closely as possible by taking advantage of the acceptance for error, approximation, explanation, and imprecise information.

8.2 THE SOFT COMPUTING – DEVELOPMENT HISTORY

Soft Computing (SC) by Zadesh in 1981

= Evolutionary Computing (EC) by Rechenberg in 1960

+ Neural Network by McCulloh in 1943

+ Fuzzy Logic (FL) by Zadeh in 1965

According to Lotfi A. Zadeh, "Smart Computing is an emerging technique that parallels the amazing capabilities of the human imagination to understand and educate in a setting of ambiguity and vagueness."

Evolutionary Computing by Rechenberg in 1960
= Genetic Programming (GP) by Koza in 1992
  + Evolution Strategies (ES) by Rechenberg in 1965
  + Evolutionary Programming (EP) by Fozel in 1962
  + Genetic Algorithms (GA) by Holland in 1970

A new interdisciplinary subject called smart computing is being developed to create the cognitive computing that is the next level of machine learning. The basic objective of smart computing is to create smart devices that can offer answers to issues of actual world that cannot be quantitatively modeled or are far too complex to use it. Its goal is to develop judgment calls that closely resemble genuine judgment by taking advantage of people's flexibility for extrapolation, ambiguity, inconsistency, and incomplete information. The characteristics of the model in this instance are approximate; they are still not exact.

Ambiguity: In this case, we cannot be assured whether the object's characteristics match those of the concept (belief).

Inaccuracy: In this case, the development model characteristics (number of variables) may always exactly be the same as the proper ones, but they are substantially similar.

Since smart computing has made it big in so many domains and is expanding quickly, its influence is likely to rise over the next few decades. Engineering and scientific research are most likely to profit the most from smart computing, yet soon its impact may be felt far beyond. In several aspects, smart computing represents an important fundamental change with goals of software development.

This shift takes into consideration the fact that, in contrast to modern computer systems, the man's brain has a great capacity for storing and processing knowledge that really is comprehensively doubtful, missing in categorization, and consisting of high technological approaches. In smart computing, the root of the issue is modeled in such a manner that allows the "condition" of the machine to be determined and measured against a desired condition. The basic principle for modifying the device's input variables, which gradually congregate toward possible answers, is still the characteristic of the software's state. The fundamental strategy used by stochastic computing and neurons computation is this. There are various ways in which smart computing is different from traditional (hard) computation. Smart computing, for instance, makes use of tolerance for imperfection, ambiguity, incomplete information, and estimation. In actuality, the subconscious person acts as a template for smart computing. The term "smart computing" refers to a group of methods from numerous domains, which fall into numerous analytical intellectual ability subcategories. Fuzzy sets, evolutionary programming, and

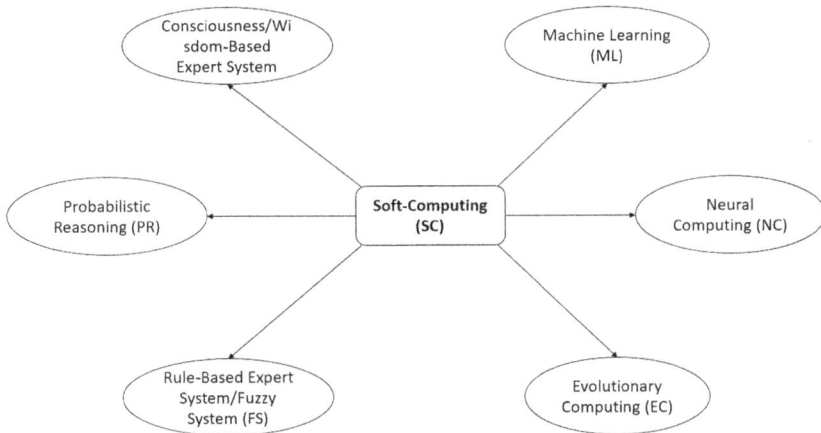

Figure 8.2 Soft-computing techniques.

adaptive neural computation are the three major subfields of smart computation. The last one includes deep learning (DL) and stochastic thinking (ST), opinion network, complexity theory, elements of observational acquisition, knowledge optimization techniques (KOT), etc., as shown in Figure 8.2.

8.3 FUZZY LOGIC (FL)

- A stochastic method called fuzzified logical enables more sophisticated judgment analysis and improved procedure-based programming incorporation.
- Conventional rationale, in where every proposition has a correct value of either 1 or 0, is developed into fuzzy systems. Assertions in fuzzy systems may have partially truth values like 0.9 or 0.5.
- Essentially, this increases the likelihood that the method will closely resemble actual situations, where declarations of unambiguous truth or falsity are uncommon.
- Statistical experts may employ a fuzzy approach to enhance the performance of various models.
- Fuzzy techniques are relatively easy to implement due to their resemblance to basic language; however, they might need careful validation and evaluation.

Unlike traditional "right versus wrong" (1 versus 0) Boolean operators developed upon which the advanced system is built, fuzzy set theory bases computation on "measures of reality." In the 1960s, LotfiZadeh at the California state University, Berkeley, became the first person to put forth the theory of fuzzy logical. Aiming to improve computational awareness of common languages was Zadeh's focus. Human language is really not translated directly through the real numbers of 0 as well as 1, similar to the majority of different events in daily existence as well as in the world [5]. The topic of which anything can eventually be described in 0 or 1 form is one worth exploring philosophically, but in actuality, most of its information we are likely to give a computer is in a stage somewhere between, as well as the outcomes of computation are typically in this form as well. It may be helpful to think of fuzzy systems as the actual process of thinking and linear, or Logical, theory as just a particular application of it shown in Figure 8.3.

It is really possible to use Fuzzy Logic, an issue-solving control system approach, in a wide range of networks, from compact, simple integrated embedded devices to massive, interconnected, number of co-PCs or desktop-based PCs, data acquisition and control operating systems. It is possible to put into practice using equipment, operating systems, or a mix of the two. Fuzzy Logic offers a straightforward method for drawing a logical decision from hazy, confusing, inaccurate, unclear, or incomplete inputs. Fuzzy Logic approaches handle difficulties similar to the way an individual should take judgments, but far more quickly [2]. A generalization of binary system, fuzzy set theory serves as the basic foundation for fuzzified models. Contrary to the binary system, although, parameters are given truth table between zero and one.

8.3.1 Fuzzy Logic in AI

Fuzzy set theory is utilized in machine learning, that is, in artificial intelligence systems to replicate human understanding and thinking. Fuzzy set theory provides zero and one as rare circumstances of validity, although having numerous intermediary levels of reality instead of simply linear instances of reality.

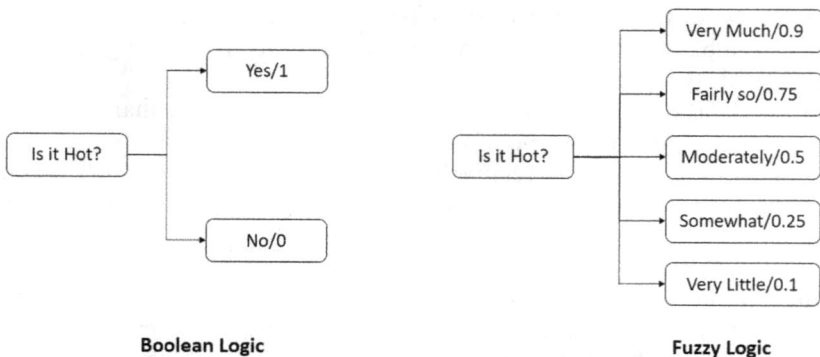

Boolean Logic   **Fuzzy Logic**

Figure 8.3 Comparing the degrees of truth with Boolean Logic vs. Fuzzy Logic.

Fuzzy set theory is therefore suitable to the aforementioned:

- Design for judgments involving ambiguous or incomplete information, like when using spoken languages understanding tools.
- Managing and adjusting equipment responses in accordance with several intakes and input factors, as in thermal controlled mechanisms.
- Perhaps the foremost potential application of fuzzy systems and fuzzy language is Indian Business Machine's Watson Super-Computer.

Under traditional collection analysis, an object's participation is simple binary, meaning that it either belongs to a given Set S or it does not. On the contrary, inside fuzzy rules, a category allows for the chance of incomplete participation. For instance, the fuzzy theory A over X is a combination of sorted pairings A = x; A(x), in that A(x) seems to be the level toward which x originally belonged to A and X = x indicates the scope of entities. If the result of the functional A(x) is 0:0, therefore, x doesn't in any way relate to A. If indeed the result is 1, therefore, x is unquestionably a constituent of A. Variables spanning 0:0 & 1:0 are used to represent an entry x partial membership in set A. So, more of x corresponds to A, closer A(x) seems to be 1:0. Fuzzy sets consequently represent a strong computer program for augmenting the capabilities of binary system in a manner which allow a far better indication of understanding in designing, for instance, an A(x) value of 0:5 indicates that x's membership inside A seems to be 50%. Figure 8.4(a) and (b) represent a membership function of fuzzy rule, which can be easily modified and its robustness is also increased.

Diagram and Block Diagram of Fuzzy Logic are given in Figures 8.5 and 8.6.

Fuzzification: The technique of transforming particular data variables into a certain level of participation of fuzzified sets depending on how closely they correspond.

Fuzzy rules/knowledge base: These would be if-then principles to abide by, which are frequently obtained through professional judgments or through better statistical methods.

Inference method: The method of arriving at the ultimate ambiguous result, taking into account the level of supply parameters membership involvement in fuzzification as well as the specific fuzzy principles.

Defuzzification: It is the procedure of transforming hazy findings into precise possible values.

8.3.2 Fuzzy set applications in power system

1. **Contingency Analysis:** Contingency or Eventuality Analysis (CA or EA) seems to be a "what if" situation emulator that ranks the effects of potential issues on an electrical PS. An eventuality is indeed the

Membership Function

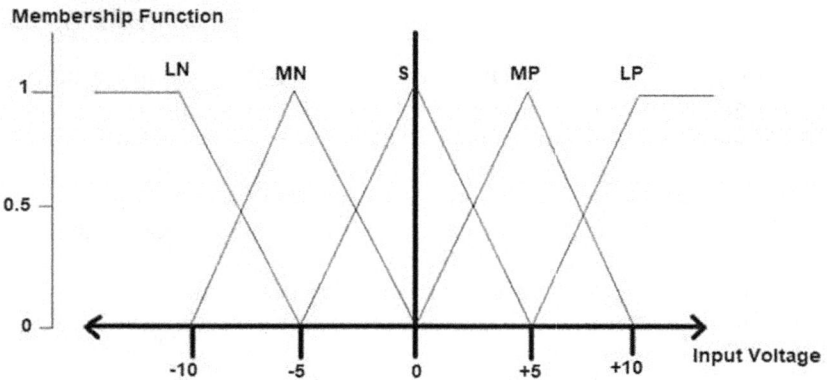

(a)

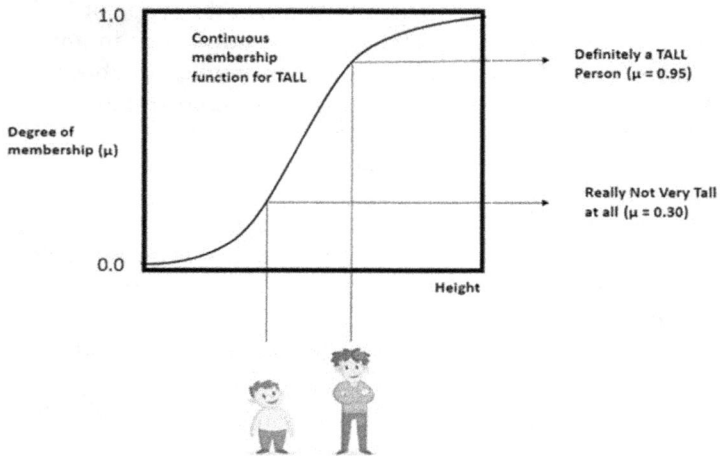

(b)

Figure 8.4 (a) and (b) Membership function.

interruption or malfunction of a specific piece of equipment, like a generating machine or transformers, or a minor component of the PS (such as a power line). This is commonly known as an unanticipated "power failure". A "forecast" analysis technique is what condition monitoring is all about. The effects of potential issues with the PS in the near future are simulated and quantified. Contingency analysis is used simultaneously as an interactive platform to demonstrate managers the consequences of potential failures and as a research instrument again for manual examination of unexpected occurrences. As a result, managers can use well before restoration simulations for being more ready to react to disruptions. A software program called CA uses a simulation analysis of the PS to assess the

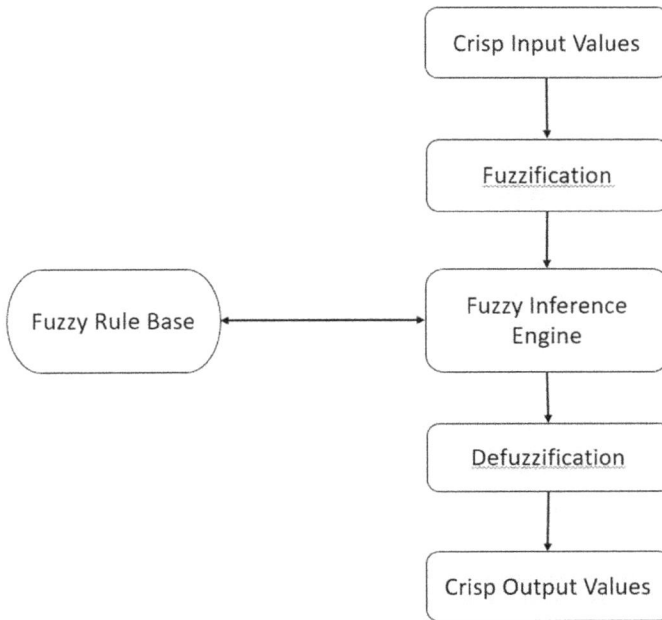

Figure 8.5 Diagram of fuzzy logic.

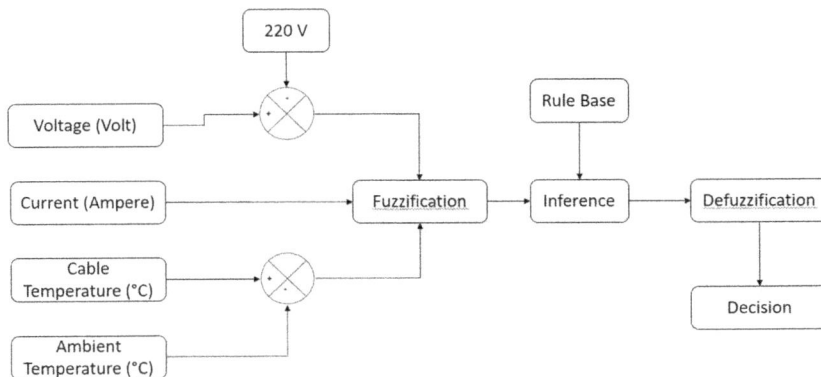

Figure 8.6 Block diagram of fuzzy logic.

- consequences of every blackout incident and
- determine probable congestions that may follow.
2. **Diagnosis/Monitoring:** Condition or State measuring (CM) is the method used to keep an eye on a different computer situation (like temp, movement, etc.) in order to spot any alterations that might be signs of an impending failure. Incorporating state monitoring is a key

component of preventive analytics since it enables the scheduling of maintenance as well as the implementation of preventative measures to stop additional breakdown and ensuing unexpected delay.

3. **Distribution Planning:** Electricity distribution planning process is a sequence of prediction, evaluation, and response preparation for electrical network preservation and growth. The objective is to operate the system's current electrical infrastructure as economically as possible while maintaining a secure, dependable, and reasonable level of support.

4. **Load Frequency Control:** An integrated Electrical system's LFC refers to the tying together of many control areas via point of common coupling. Both change in frequencies and tie – line divergence are caused by sudden load variations by almost any regulatory region of a coupled electricity system.

5. **Generator Maintenance Scheduling:** Developing a timetable that allows for such scheduled repairs to be carried out on the generators in an electricity system is known as the generators management schedule (GMS) challenge.

6. **Generator Dispatch:** When we talk about distributed power, then we are talking about renewable electricity, which electricity supply managers may schedule as needed to meet market demands. Distributed power units are capable of altering their energy production in response to a command.

7. **Load Flow Computations:** Under a specific load and generation of real electricity and potential circumstances, the goal of electricity flow analysis computations will be able to identify actual stable operation of an electricity system. With such a knowledge, researchers can quickly and simply determine the P and Q power flow in every section as well as corresponding power dissipation.

8. **Load Forecasting:** By foreseeing forthcoming demand of electricity that the power company will transfer or distribute, demand forecasts reduces utilities risk. Pricing, climate and market feedback evaluation, and prediction for renewable electricity are examples of methods.

9. **Load Management:** By modifying or monitoring the consumption instead of just overall power plant production, demand monitoring, commonly referred to as requirement planning (Demand Side Management), comprises of the technique of regulating the electricity that is available here on system more with the utility grid.

10. **Reactive Power/Voltage Control:** Voltage regulation and Q-power control are separate facets of a same task that boosts dependability and makes it easier for businesses to trade with one another through electric grid. Output voltage gets managed around an AC current system through controlling the generation and consumption of Q-power.

11. **Security Assessment:** A system's ability to ensure ongoing functioning under normal circumstances, irrespective if certain scenarios arise, is referred to as electrical system security. According to such concept, an electrical network's security evaluation considers when an electric grid operational points can endure various plausible and legitimate scenarios.

12. **Unit Commitment:** A common issue inside the electrical network is unit commitments (UC), which seeks to reduce the overall expense of electricity production over a certain time duration by creating an appropriate schedule for the production units. Numerous operating restrictions must be honored by the UC approach [21].

8.3.3 Fuzzy Logic applications in other fields

Fuzzy set theory is used in a variety of artificial intelligence systems and technologies. These encompass technology products, computing, chemistry, pharmaceuticals, and aviation in addition to automobile intelligence or knowledge in Figure 8.7.

- Fuzzy set theory is employed in cars to choose the latest gear depending on variables, including cylinder pressure, traffic situations, and riding attitude.
- Fuzzy set theory in dishwashing uses elements like the quantity of meals and the volume of nutrients residues here on dinnerware to calculate the cleaning technique and power consumption required.
- Fuzzy set theory is employed in photocopiers to modify the drum energy depending on variables including climate, moisture, and image intensity.
- Fuzzy set theory is employed in aviation to adjust satellites and rocket height depending on external inputs.
- Fuzzy set theory is employed in healthcare to make computer-assisted diagnosis depending on elements like indications and health information.
- Fuzzy set theory is employed in pharmaceutical evaporation to regulate temperatures and acidity factors.
- Fuzzy set theory is employed in spoken language translation to establish conceptual relationships among ideas expressed by speech as well as additional output factor.
- Fuzzy set theory makes outcome decisions for climatic monitoring devices like air conditioning units and warmers dependent on variables like the ambient temperatures and the desired temperatures.
- Fuzzy set theory can be utilized within a corporate standards system to speed up judgment based on specified conditions.

What are the uses of Fuzzy Logic?

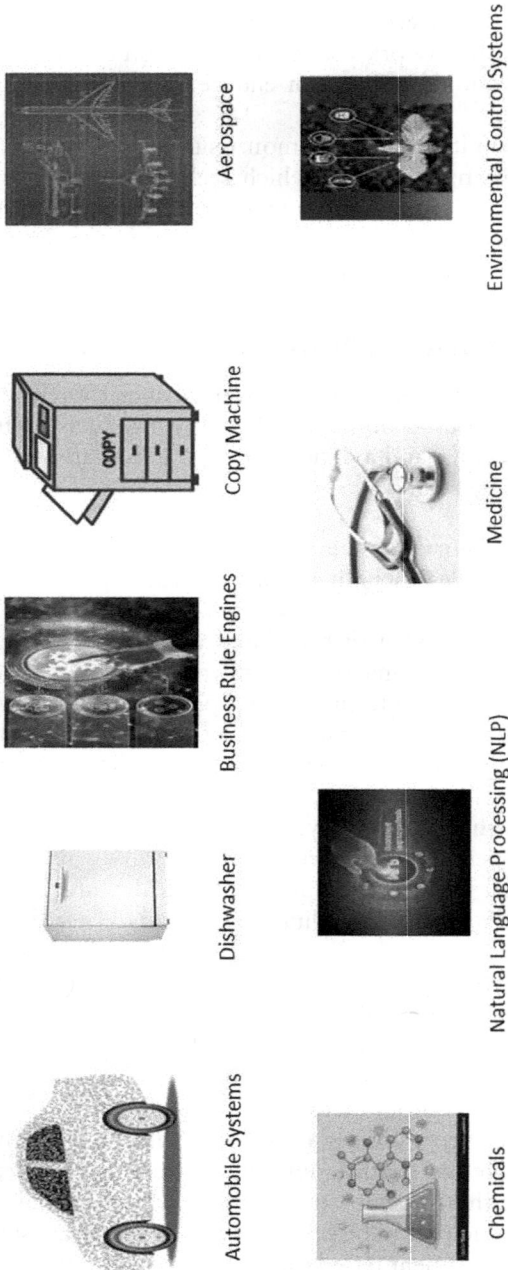

Aerospace

Environmental Control Systems

Copy Machine

Medicine

Business Rule Engines

Natural Language Processing (NLP)

Dishwasher

Automobile Systems

Chemicals

Figure 8.7 Fuzzy logic uses in technology.

8.4 ARTIFICIAL NEURAL NETWORK (ANN)

The artificial neuron structure (ANN) is really a framework for information transfer that takes its cues from how human neuronal networks, like the brains, function. The integral member of such data handling mechanism is the fundamental component of such a methodology. It is made up of several, intricately linked functioning units called neurons that collaborate to address particular issues. ANNs develop via imitation just like humans do. Throughout an instructional cycle, an ANN is tailored for a particular purpose, including as analytical thinking or data preprocessing. Modifications on the neural interconnections, which occur across two neural relationships, are a result of training in living organisms. It also applies to ANNs [5]. Several ways that each neural circuit might cluster affect how neural network model functions. In the biosphere, minute parts are assembled into three-dimensional neuronal networks in Figure 8.8. Such nerves appear to have almost limitless connectivity potential. In a particular instance of either suggested or actual man-made link, this is not accurate. With present technologies, embedded circuitry constitutes two-dimensional objects despite a finite degree of interconnecting regions. The kinds and range of artificial brain circuits, which might be executed in semiconductors, are constrained by such actual fact. At the moment, human brains are only a simple grouping of artificially rudimentary brains. By building layers and connecting them, such clusters take place. The major aspect of "arts" of designing systems to fix issues within the actual world is when such levels interact [6].

Although algorithms are created to imitate both humans and animal's minds, neuron network models set themselves apart among other types of synthetic thinking. Fake neurons are networks of interconnected networks that collect, analyze, and transfer information. To decide when data inputs surpass a specific barrier, every node applies a computational calculation for the data information obtained [7].

The neurons activate and every information moves onto the following layers whenever the outcome surpasses the specified limit. Such networks may be programmed to continuously alter parameters in such a manner that yields an expected outcomes utilizing a number of educational methods. Several levels of ML (Machine-Learning) algorithms [8] successively retrieve relatively high characteristics using incoming datasets using a deeply studying framework. Such techniques advance beyond recognizing general characteristics in datasets, pictures, or sound to noticing minute features. A transportable gadget, which continuously transforms whole phrases of American Signing Speech into other voiced English, is one example of how automation of such jobs offers dramatic advances in a number of areas.

Probabilistic gradient descent is a technique used by several deep intelligence systems to optimize variables. By using a back propagation training technique to calculate the gradients of damage units, a neuron network may be trained to select appropriate categories from the data obtained during a

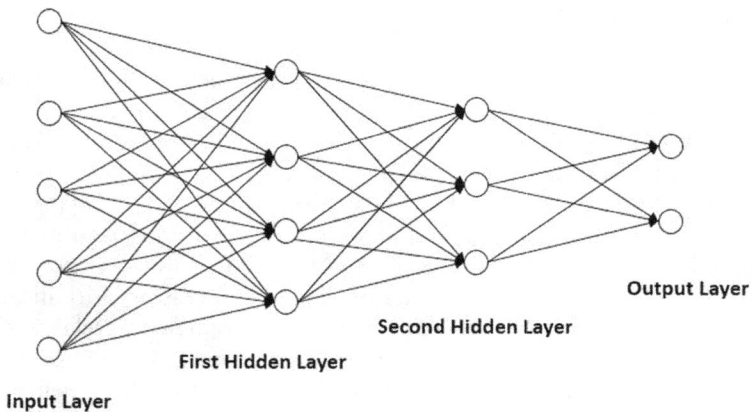

Figure 8.8 Feed forward multi-layer perception network.

guided operation of training. In other words, the technology assesses how effectively a program predicts the pertinent facts and makes adjustments as necessary. A machine might train to carry out activities like identifying a certain sort of thing in such a collection of pictures or identifying the phrases in such a recording of human's voice during the optimal solution procedure. Although supervised training has drawn a lot of awareness, another intriguing method of creating neuron systems is neurological development. Rather than using diffusion-based optimization, the method uses evolution-based algorithms to mimic how living species' brains evolved biologically. The procedure begins from neurons that are generated from randomness and are probably all unable to carry out simple tasks. The most prosperous individuals are chosen to spread their achievement toward the following generations. Such human brains produce progressively adept descendants through successive repeats that are capable of handling challenging tasks like operating robotics or participating in computer games. Despite the fact that neurons may mimic the mind activity, researchers and technologists still have a long way to go until machines may mimic humans understanding. By modeling how creatures developed the capacity for learning, experts from the disciplines of biological sciences, data science, and technology worked together to assist us comprehend the difficulties encountered. A manmade neuron, like the scientists noted, does have a significant advantage beyond methods developed via organic mutations since the manmade synapses are formed with the intention of forming linkages since the beginning [9].

8.4.1 Electrical engineering and machine learning applications for neural networks

Effective methods and techniques for electrical and electronics engineering have been developed using machine learning as well as other types of AI technology. Fuzzy logic-based controller systems, which can take into consideration a continuity of circumstances while developing guidelines for just how devices react to stimuli, are one illustration of the how intelligence is being used in practice today. Scholars are examining whether artificial neural-based cloud might provide additional benefits for managing power stations and enhancing functionality in nonlinear systems. The efficiency and reliability of signals recognition and treatment are being improved through ML and artificial intelligence. Programs were employed, for instance, to analyze impulses, extract information from signal inputs, and modify output voltages. Research also indicated the possibility of improving numerous signal recognition and manipulation by building neurostructures [10]. Probabilistic signals might be applied by deep and wide learning techniques to lower the interference that might sometimes obstruct speech augmentation or picture identification.

The reliable production of sustainable power may potentially benefit from deep learning. The fact is that sunlight does not constantly shine as well as the air does not constantly flow is probably one of the biggest drawbacks of depending on renewable energy resources such as solar and wind. Because of this, planning the power grid is essential to ensuring a steady source. Forecasting can be made more effective by using a computational model to quickly evaluate a vast proportion of information and create better precise forecasts of wind velocity. A telemetry surveillance method can record time frames as to microseconds while noting temp, wind velocity, wind patterns, and atmospheric pressures by using a multivariate system. The development of neurons offers researchers countless chances for experiments and original dilemma. Experts might discover innovative methods to use machine learning and machine intellectual ability to accomplish goals by developing their electric power and software technology skills. Scholars can design complex learning models and carry out deeper learning tasks, thanks to the coursework in State Universities, which provide Master of Technology in Electrical and Computing Technology programs. General Classification is given in Figure 8.9. This program assists electrical and computing professionals in acquiring the cutting-edge professional competencies needed to power the future iteration of technologies in industries like medicine, sustainable sources, and driverless vehicles. From 1988, when they were first used in the field of electrical networks, ANNs had been used to numerous different uses. This field of study is quickly picking up steam. The majority of the material cited inside this work, greater than 75% of the total, was released within the previous 20 months [11]. The majority of implementations are found in fields in which historic evidence must be used to classify patterns. The authors report the techniques listed as well as the neural models employed in the following:

- Hopfield system, multi-surface convolution, web-based security evaluation.
- Imagination features mapping, multi-surface feed-forward neural, demand prediction.

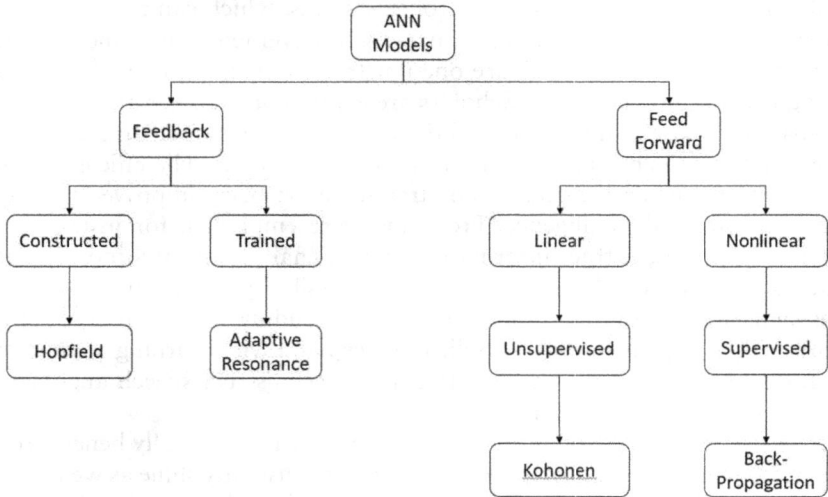

Figure 8.9 General classification of ANN models.

- Multilayered convolutions, fault localization.
- Multilayered convolutional layer for high susceptibility defect monitoring.
- Multilayered convolutions for early induction motor drive defect detection.
- Interpretation of alarms, multilayered perceptron's optimum transitioning of circuitry and multilayered perceptron's.
- Multi-layered Convolutions with Power Network Stabilization Adjustment.
- Surveillance and management of the furnace, multilayered perceptron's.
- The study of eddy currents and neuronal artificial network.
- Multilayered convolutions, sinusoidal origin recognition.
- Examination of hierarchical visibility, multilayered perceptron.
- Hierarchical perceptron, adaptable sequential combiner, multilayered perceptron.
- Hopfield networks, multilayered neural network model, and hazard filtering.
- Nilsson's thinking system, multilayered perceptron, or synchronous motor modelling and evaluation.
- Multilayered convolutions, power conversion controller.
- The multilayered convolutions, motor drive controls.
- Evaluation of dynamic response, multilayered convolution.
- Multilayered convolutions, regulation management of the turbo generator.
- Unit commitment, multilayer perceptron.
- Strategy for growth, multilayered perceptron's.

- Surveillance and diagnostics of nuclear energy plants, multilayered perceptron's and equality space models.
- Multilayered convolutions, improvement of the turbines back pressure.

The largest widely used ANN technique is probably swing inspired convolution simulation. Further solutions include enhanced Hopfield systems, personality convolution layers, and mobile connections [12]. Using a careful assessment and remark, the above article summarizes the significant implications that have been discussed.

8.4.2 Engineering applications of neural network

Scientists can benefit greatly from computer designs that become adept at carrying out necessary functions or keeping an eye open for irregularities. Experts have shown the adaptability of this type of machine learning in a variety of initiatives, a few of them might have profound effects upon the development of medicine and shipping:

- Perhaps the biggest important sectors for applying depth understanding is driverless cars. Driverless cars can collect sensed details from the highway forward, respond to impediments, but also execute judgments like travel routing thanks to neurons.
- A CANVAS Mind is a deeper knowledge architecture being developed by scholars in MSU's Connecting Automatic Integrated Cars and trucks for Proactive Security department that uses sensory synthesis to comprehend the environment. This method combines information via detectors, lasers, plus hyperspectral camera systems to offer special monitoring, which can allow automated automobiles make judgments in true or alert humans motorists to dangerous road surfaces.
- In order to improve the efficiency of vehicles, automobile researchers have suggested further use for neurons in automobiles, such as determining the physiological and biochemical characteristics of different fuel kinds.
- Accident prevention techniques are among the several neurosystem applications, which might be used on aeroplanes, and they might result into better and greater effective performances. According to another research, a multilayer recurring neuronal system could forecast 4D trajectories with high accuracy while taking weather conditions like winds, temperatures, and thunderstorms into consideration.
- Professionals in computing algorithms are looking into methods to enhance the quality of medical services and services. Examples of such ideas involve anticipating the frequency of elderly patients' health facility visits, spotting earlier Brain cancer illness symptoms in MRI tests, even analyzing medical records more quickly using technologies, which may decipher acronyms.

8.5 ADAPTIVE NEURO FUZZY INFERENCE SYSTEM (ANFIS)

Using densely integrated artificial neuro computing components and informational linkages that also are graded to translate various quantitative entries in and out of an outcome, ANFIS is indeed a straightforward dataset acquisition method that employs fuzzy set theory to convert specified values into a desirable outcome. ANFIS integrates the advantages of fuzzified thinking and neuronal networks, two artificial intelligence methodologies, into the single strategy. This ANFIS operates by using artificial neural modeling techniques to adjust the fuzzy rule program's variables (FIS) [13]. These are a number of characteristics, which help ANFIS succeed. In order to explain the behavior of a complicated network, it improves fuzzified IF-THEN constraints.

- No additional human skill is necessary.
- Implementation is simple.
- It makes understanding quick and precise possible.
- It provides the needed set of data.
- More memberships feature to choose from.
- Great ability to generalize.
- Fantastic explanatory tools using fuzzy set; and
- When fixing difficulties, it is simple to combine verbal and mathematical information.

Fuzzy intelligence algorithm and neuron networks are combined inside an adaptable neurological fuzzy approach. Tsukamoto and Sugeno fuzzy frameworks can indeed be described either by ANFIS framework or by estimating the variables in ANFIS. This ANFIS model functionality is similar to that of the directional-based functional networks (DBFN), although having modest restrictions. A combination framework of fuzzy-based and neuronal networks technology, which makes up such ANFIS approach, is given in Figure 8.10. Whereas the neuron lends the technology a perception of flexibility, fuzzy set theory accounts for the program's inaccuracy and uncertainty. Through the assistance of the constraints deduced from the inputs and expected output data of the systems getting modeled, a basic fuzzy controller as well as its input parameters are initially constructed utilizing such mixed approach. This ultimate ANFIS version of this framework then is created by fine-tuning existing constraints of said basic fuzzified structure using the neuro networks. An automated neuron system, which is founded on the Takagi-Sugeno fuzzification process, is known as an adaptable neuro-fuzzy intelligence strategy (ANFIS) or adaptable network-based fuzzy method of assessment [14]. In the earlier 1990s, we saw the advancement of the method. It possesses the ability to combine the advantages of neuronal systems and fuzzified processing in a unified platform because it blends each of these concepts [15]. Its own interpretation mechanism

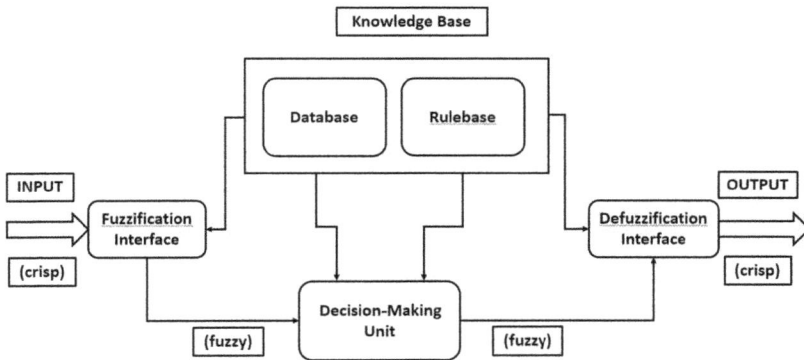

Figure 8.10 Fuzzy inference system.

is a collection of IF-THEN fuzzy set theory with the flexibility to understand and estimate complex functions. As a result, ANFIS is regarded as a global estimation [16]. The optimum characteristics discovered through an evolutionary algorithm can be used to utilize ANFIS better effectively and efficiently. It can be used in sophisticated energy planning systems that are situationally aware.

8.5.1 Blocks of FIS

- A set of fuzzification if-then instructions in a decision table.
- A collection that contains information about the transfer mechanism of the intuitionistic fuzzy utilized in the fuzzification.
- A judgment unit that applies those laws of inference processes.
- A fuzzified mechanism that converts the sharp input variables into levels of linguistics value compatibility.
- A defuzzified mechanism that turns an interface's hazy outputs in a clear outcome.

8.5.2 Steps of fuzzy reasoning

1. To determine every linguistics designer's inclusion algorithm (or consistency measurements), evaluate the supply factors with the decision variables in its premises component. This process is frequently known as fuzzing.
2. These memberships variables on the premises part should be combined (using a particular T operation, typically multiplier or min.) to obtain the discharge intensity (weighted) of individual rule.
3. Based on the discharge intensity, produce a certified consequence (fuzzification or crisper) of every rule.

4. Combine the appropriate results to get precise results. This procedure is known as defuzzification.

8.5.3 Types of fuzzy reasoning

Type 1:   The total outcome is the calculated mean of the outcome participation and discharging intensity of individual rule.

Type 2:   Implementing the "max" technique to the certified fuzzy outcomes results in the cumulative output. Several defuzzification techniques can be used to produce the finished, proper result.

Type 3:   Fuzzy if-then procedures by Takegi and Sugeno are employed. This resulting outcome is the balanced mean of the outputs from all the guidelines combined. Every decision's outcome is a proportional composition of input data and as well as a fixed term.

ANFIS, which Jang successfully created, combines ANNs and fuzzification systems (FIS). It is a multidimensional input network that maps an incoming region to an outcome space using fuzzified thinking and machine learning based training techniques. The memberships functions (MFs) are used to construct the fuzzified judgement procedures, as well as the system develops the MFs' strongest variables. Every position in the intake region is projected to a memberships values (or level of memberships) within 0 and 1 using a curves known as a MF [17]. FIS originally built on if-then procedures, which allow us to use them to determine the relationship among both incoming and outcome values. We employ FIS as a predictive framework since the traditional estimating methods, such as extrapolation, do not adequately account for dataset ambiguity in situations when both input and outcome variables has high levels of ambiguity.

This ANFIS is indeed a statistics method for solving functions approximated issues that uses neural networks. The conventional statistics method for creating ANFIS systems is dependent on aggregating trained batches of mathematical values of the arbitrary functions that needs to be estimated. ANFIS circuits had also been effectively used since their invention for categorization jobs, regulation processes controls, pattern matching, and other issues of a similar nature [18]. Here, a fuzzification inference system is made up of a fuzzification that Takagi, Sugeno, and Kang developed in order to define a methodical process for generating fuzzy sets from such an interface dataset. ANFIS, a composite model made up of a fuzzified and artificially neuron, tries to solve engineering issues by predicting the behavior of inaccurately complicated dynamical system. ANFIS employs a hybrid method blending gradients reduction, backward dispersion, and a minimum approach to generate a range of variables and contains an adjustable ANN as well as a fuzzy approach for inference. ANFIS are sometimes viewed as a flexible mathematics architecture that really can compute-level forecast a wide range of complicated highly nonlinear network with the required extent of accuracy.

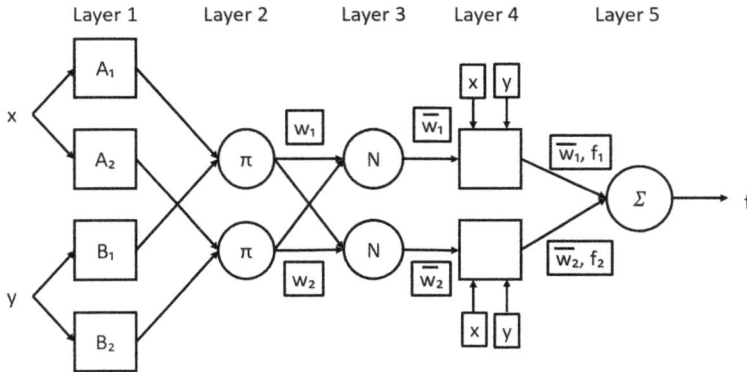

Figure 8.11 Structure of ANFIS.

Five phases make up the ANFIS architecture: a fuzzified layer, a products layer, a normalized layer, a defuzzified layer, and an overall production layer, which is given in Figure 8.11.

- **Layer 1**: Each component in this tier is an adapted element. Terminal procedures include the generalized ring degree of membership and the Gaussian memberships variable.
- **Layer 2**: Every node outcome in this level displays the discharge rate of a rules.
- **Layer 3**: The normalized discharge intensity of all rules is represented by individual node.
- **Layer 4**: Each node in this level is adaptable and has a network function that describes how the regulations contributed toward the final outcome. Consequential variables are the name given to the variables in these levels.
- **Layer 5**: The total of each rule outputs is computed by a singular node.

Throughout this research, various ANFIS models have been developed utilizing instructional data from the eight cities and seven input variables to estimate GSR. The outcomes of such ANFIS systems have been then contrasted using information collected during experimental years. The fuzzified intelligence systems underneath discussion comprises of two input parameters and single outputs, which is considered for convenience. The fuzzified if-then principles of Takagu and Sugeno's kind are contained inside the knowledge base:

If x is A and y is B then z is $f(x,y)$

Suppose the fuzzified sets within the antecedent variables are A & B and $z = f(x, y)$ is an excellent component in the result Normally, for the inputs elements x and y, $z = f(x, y)$ is a polynomials. However, it may simply be anything other variable that may roughly characterize the software's outputs in the hazy area defined by its antecedents. A no level Sugenofuzzified system, which might well be thought of being a specific example of the Mamdanifuzzified judgment systems whereby each rules subsequent is described by a fuzzified singularity, is created where $f(x, y)$ is a consistent. A initial level Sugenofuzzified models is created if is assumed to be $f(x, y)$. The two rules for the first-level Sugeno fuzzified inferences systems can be written as [20]

Rule 1:
If x is A_1 and y is B_1 then $f_1 = p_1 x + q_1 y + r_1$

Rule 2:
If x is A_2 and y is B_2 then $f_2 = p_2 x + q_2 y + r_2$

Therefore, the Takagi and Sugeno type-3 fuzzy linguistic method is utilized. Every legislation's outcome in such reasoning method is a proportional mixture of the inputs plus a continual component. The weighed mean of the results from each criterion serves as the ultimate outcome. The exact matching ANFIS architecture is shown in Figure 8.1.

The individual layers for this ANFIS structure is described below:

- **Layer 1:** Every node i in this layer is adaptive with anode function

$$O_i^1 = \mu_{Ai}(x)$$

where x is the input to node I, A_i is the linguistic variable associated with this node function, and μ_{Ai} is the membership function of A_i. Usually, $\mu_{Ai}(x)$ is chosen as

$$\mu_{Ai}(x) = \frac{1}{1 + \left[\left(x - \frac{c_i}{a_i}\right)^2\right]^{b_i}}$$

Or

$$\mu_{Ai}(x) = \exp\left\{-\left(c - \frac{c_i}{a_i}\right)^2\right\}$$

where x is the input and $\{a_i, b_i, c_i\}$ is the premise parameter set [6].

- **Layer 2:** Each node in this layer is a fixed node, which calculated the firing strength ω_i of a rule. The output of each node is the product of all the incoming signals to it and is given by,

$$O_i^2 = w_i = \mu_{Ai}(x) \times \mu_{Bi}(y), i = 1, 2$$

- **Layer 3:** Every node in this layer is a fixed node. Each ith node calculated the ratio of ith rule's firing strength to the sum of firing strengths of all the rules. The output from the ith node is the normalized firing strength given by,

$$O_i^3 = \varpi_i = \frac{w_i}{w_1 + w_2}, i = 1, 2$$

- **Layer 4:** Every node in this layer is an adaptive node with a node function given by [18]

$$O_i^4 = \varpi_i f_i = \varpi_i \left(p_i x + q_i y + r_i \right), i = 1, 2$$

where ϖ_i is the output of Layer 3 and $\{p_i, q_i, r_i\}$ is the consequent parameter set.
- **Layer 5:** This layer comprises of only one fixed node that calculates the overall output as the summation of all incoming signals [19], that is,

$$O_i^5 = \text{overall output} = \Sigma_i \varpi_i f_i = \frac{\Sigma_i w_i f_i}{\Sigma_i w_i}$$

8.5.4 Application areas of ANFIS

The applications of ANFIS are mentioned above in the Fuzzy Logic System and Neural Network System already. The ANFIS system is used due to certain advantages over other methods. The mentioned key fields for ANFIS application domains are highlighted: robotics, autonomous control, patterns recognizing, unbalanced forecasting, dynamic systems verification, and adaptable signals analysis. ANFIS is used to develop two ill-defined structures, including indoor thermal as well as collective innovations in manufacture and operational activities planning [9], surface temp supervision of a water hot tub mechanism, monitoring of certain machining procedures, cable electrical release machining, milling, equipment monitoring, and uninteresting procedures, equipment situation tracking.

- To prevent thermal deformation to the item as well as the equipment during cutting operations like pulverizing, sawing, twisting, digging, and jigs digging.
- To prevent tool breakdowns, chopping powers are controlled during milling operations like sawing, grinding, and twisting, for instance.
- When using machining techniques like cutting and grinding, it is important to regulate the dimensions and geometry of a piece.
- In lasers cutting, when focusing a lasers stream.
- High-voltage modification in electric discharges milling.
- Reducing disturbances during machining operations.
- Waters jet concentrating during water jet milling.
- Automatically generated machining processes automatically supply and arrange components.

8.6 CONCLUSION

The various soft computing techniques are discussed in this chapter. The wide applications of these all soft computing in power system are discussed. Fuzzy techniques are relatively easy to implement due to their resemblance to natural languages, although also may need careful analysis as well as evaluation. An ANN represents a framework for information transfer that takes its cues from how neurological systems, like the mind, function. The innovative architecture of the systems that process knowledge is the fundamental component of this approach. The ANFIS combines the benefits of Fuzzy Systems (FL) and Neuron Network (ANNs) into a single architecture. In order to model complicated patterns and comprehend nonlinear interactions, it offers rapid learning performance and flexible analytic skills. ANFIS is being utilized and put into effect in many different fields, offering answers to persistent issues with reduced both time and location complication.

REFERENCES

1. IEEE Task Force on Blackout Experience, "Mitigation, and role of new technologies, blackout experiences and lessons, best practices for system dynamic performance, and the role of new technologies", Technical Report. Special Publication 07TP190, IEEE (2007).
2. P. Kundur, *Power System Stability and Control*, Mc Graw Hill, New York, 1994. "Tools and industry experience", IEEE Committee Vol. IEEE/PES 93TH0358-2-PWR 19.
3. Lavanya Neerugattu, et al., "New criteria for voltage stability evaluation in interconnected power system", In *National Power Systems Conference* Corpus ID: 102344925 (2012).
4. G. Naadimuthu, D. M. Liu, E. S. Lee, "Application of an adaptive neural fuzzy inference system to thermal comfort and group technology problems",

Computers & Mathematics with Applications, 54, 11–12, December 2007, 1395–1402.

5. W. Wang, L. Qiu. "Prediction of annual runoff using adaptive network based fuzzy inference system", In *2010 Seventh International Conference on Fuzzy Systems and Knowledge Discovery* (pp. 1324–1327). IEEE (2010) .

6. H. Khayyam, S. D. SaeidNahavandi, "Adaptive cruise control look-ahead system for energy management of vehicles", *Expert Systems with Applications*, 39, 3, 15 February 2012, 3874–3885.

7. P. Kayal, S. Chanda, C. K. Chanda, "Determination of voltage stabilityin distribution network using ANN technique", *International Journal on Electrical Engineering and Informatics*, 4, 2, 2012, 347.

8. W. M. Villa-Acevedo, J. M. López-Lezama, D. G. Colomé, "Voltage stability margin index estimation using a hybrid kernel extreme learning machine approach", *Energies*, 13, 4, 2020, 857.

9. Syed Mohammad Ashraf, et al., "Voltage stability monitoring of power systems using reduced network and artificial neural network", *International Journal of Electrical Power & Energy Systems*, 87, 2017, 43–51.

10. S. K. Tiwary, J. Pal, "ANN application for voltage security assessment of a large test bus system: a case study on IEEE 57 bus system", In *2017 6th International Conference on Computer Applications in Electrical Engineering-Recent Advances (CERA)* (pp. 332–334). IEEE (October 2017).

11. Z. G. Sanchez, et al., "Voltage Collapse point evaluation considering the load dependence in a power system stability problem", *International Journal of Electrical & Computer Engineering*, 10, 1, 2020, 61–70.

12. M. Moghavvemi, F. M. Omar, "Technique for contingency monitoring and voltage collapse prediction", *IEEE Proceeding on Generation, Transmission and Distribution*, 145, 6, November 1998, 634–640.

13. M. V. Suganyadevia, C. K. Babulalb, "Estimating of loadability margin of a power system by comparing Voltage Stability Indices", In *2009 International Conference on Control, Automation, Communication and Energy Conservation* (pp. 1–4). IEEE (2009).

14. I. Musirin, T. K. A. Rahman, "Novel fast voltage stability index (FVSI) for voltage stability analysis in power transmission system", In *2002 Student Conference on Research and Development Proceedings*, IEEE, Shah Alam, Malasia (July 2002).

15. K. Arora, A. Kumar, V. K. Kamboj, D. Prashar, B. Shrestha, G. P. Joshi, "Impact of renewable energy sources into multi area multi-source load frequency control of interrelated power system", *Mathematics*, 9, 2021, 186.

16. K. Arora, A. Kumar, V. K. Kamboj, D. Prashar, S. Jha, B. Shrestha, G. P. Joshi, "Optimization methodologies and testing on standard benchmark functions of load frequency control for interconnected multi area power system in smart grids," *Mathematics*, 8, 2020, 980.

17. J.-S. R. Jang, "ANFIS: adaptive-network-based fuzzy inference system", *IEEE Trans Systems Man Cybern*, 23, 3, 1993, 665–685.

18. P. Singh, K. Arora, U. C. Rathore, "Control strategies for improvement of power quality in grid connected variable speed WECS with DFIG–an overview", *Journal of Physics: Conference Series*, 2327, 2022, 1–14.

19. J.-S. Jang, C. T. Sun, E. Mizutani, *Neuro-Fuzzy and Soft Computing*, Prentice-Hall, Englewood Cliffs, NJ, 1997.

20. M.-Z. Huang, J.-Q. Wan, Y.-W. Ma, W.-J. Li, X.-F. Sun, Y. Wan, "A fast predicting neural fuzzy model for on-line estimation of nutrient dynamics in an anoxic/oxic process", *Bioresource Technology*, 101, 6, 2010, 1642–1651.
21. K. Arora, *Internet of Things-Based Modernization of Smart Electrical Grid, Smart Electrical Grid System*, CRC Press, Routledge, Taylor & Francis, Boca Raton, 2022.

Chapter 9

Maize diseases diagnosis based on computer intelligence

A systematic review

A. Dash and P. K. Sethy
Sambalpur University, Burla, India

CONTENTS

9.1 INTRODUCTION

It is hard to imagine living without agriculture, just as breathing without air is impossible. Agriculture not only provides food but also impacts jobs and income globally. The agricultural sector is vital to many countries' economic development. Agriculture is also critical to the economic development of many countries, including India, China, the United States, Brazil, Mexico, Russia, Japan, Germany, and France. As reported by the Food and Agriculture Organization (FAO), agriculture employs 67% of the total population and accounts for 39.4% of GDP, with agriculture goods accounting for 43% of all exports [1]. Figure 9.1 shows the top 20 countries globally based on the contribution of agriculture to their nation's GDP.

According to TheGlobalEconomy [2], the GDP share of agriculture in India for 2020 is 18.32% and stands at 40th place in global ranking.

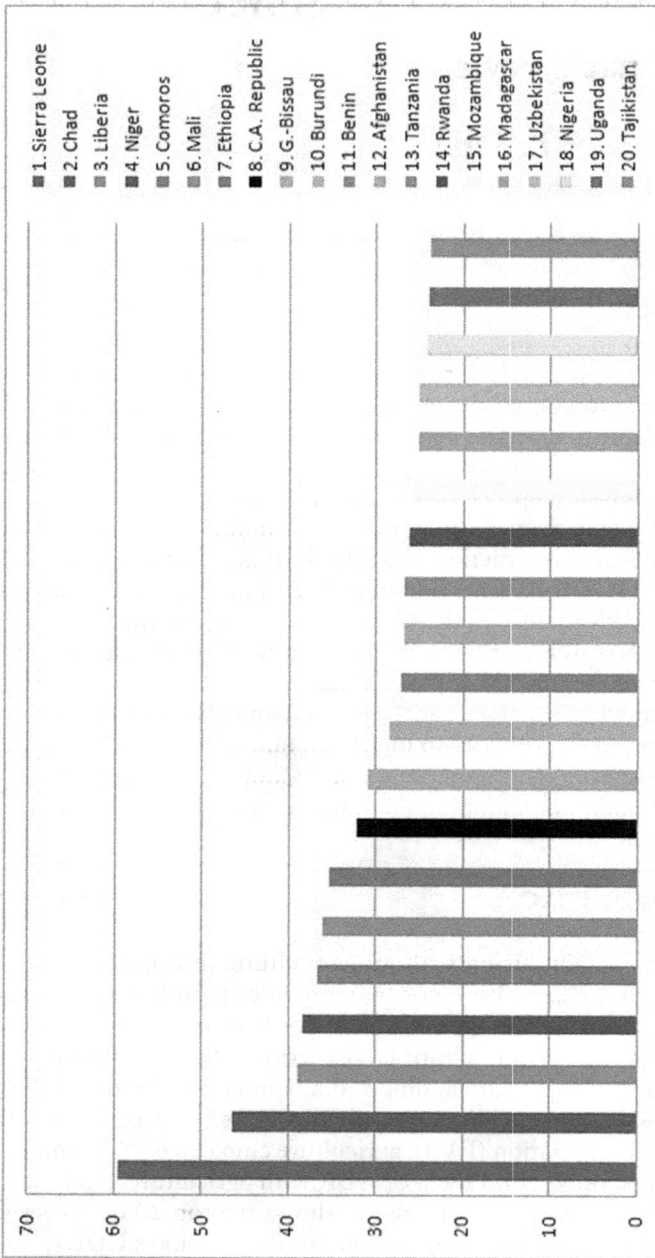

Figure 9.1 Top 20 countries based on contribution of agriculture to their GDP [2].

Figure 9.2(a) shows the year-wise contribution of agriculture to the national GPD from 2013 to 2020. However, India is already lagging in terms of domestic food production due to a variety of factors such as an unplanned management, uncertain weather conditions, and inefficient harvesting and improper irrigation techniques, to name a few [3]. Therefore, it is essential to increase the study and analysis of the technical methodologies that can improve productivity and the quality of different agricultural products.

Maize (Zea mays L) is one of the most adaptable emerging crops, accounting for 36% (782 million tons) of global grain production and roughly 9% of India's national food basket. In countries like India, maize is the third most significant food crop along with wheat and rice. Mostly, this crop is sown in two seasons: rainy (Kharif) and winter (Rabi). It is utilized in hundreds of industrial products and industries such as beverages, starch, sweeteners, oil, protein, pharmaceuticals, cosmetics, film, textile, gum, and packaging. According to the ICAR-Indian Institute of Maize Re-search [5], India ranks fourth in terms of area and seventh in terms of production, representing nearly 4% of global maize acreage and 2% of the total output. Figure 9.2(b) shows that approximately 47% of maize production is used for chicken feed, 14% for starch, 13% for cow feed, and 7% and 6% for processed food and export, respectively. The remaining 13% is utilized for livestock feed and food.

Maize infections are among the most significant impediments to maize production, although they can be addressed with early detection. But the farmers face substantial economic losses due to many causes like climate change, insect attacks like shoofly and chaffer beetle, and different types of bacterial and fungal diseases in several parts of maize plants, which can be seen from leaf to panicle. But in general, the detection and classification of maize diseases are done by some agricultural experts or by the farmer themselves based on their experience, which is inefficient and time-consuming. Identification and classification of the maize plant diseases with the help of computer intelligence (image recognition based on ML and DL methodologies) can be one of the efficient solutions to the above problem. This chapter represents a detailed review of the different kinds of maize diseases and the approaches based on computational intelligence to identify and classify these diseases.

In this article, 156 research papers are studied and analyzed, published between 1999 and 2022 in various publications and conference proceedings to identify and classify maize and other plants diseases. From these 156 numbers of research papers, we selected 81 numbers for our review, which is detailed in the following parts. Figure 9.3 depicts the year-by-year published research papers selected for our review study.

The remnant of the work is organized as follows. Section 9.2 illustrates the various maize plant diseases. Section 9.3 covers basic approaches and ML and DL-based classifiers. In Sections 9.4 and 9.5, we discuss recent research on the disease classification of maize and other plants. Eventually, we draw conclusion and summarize the review work in Section 9.6.

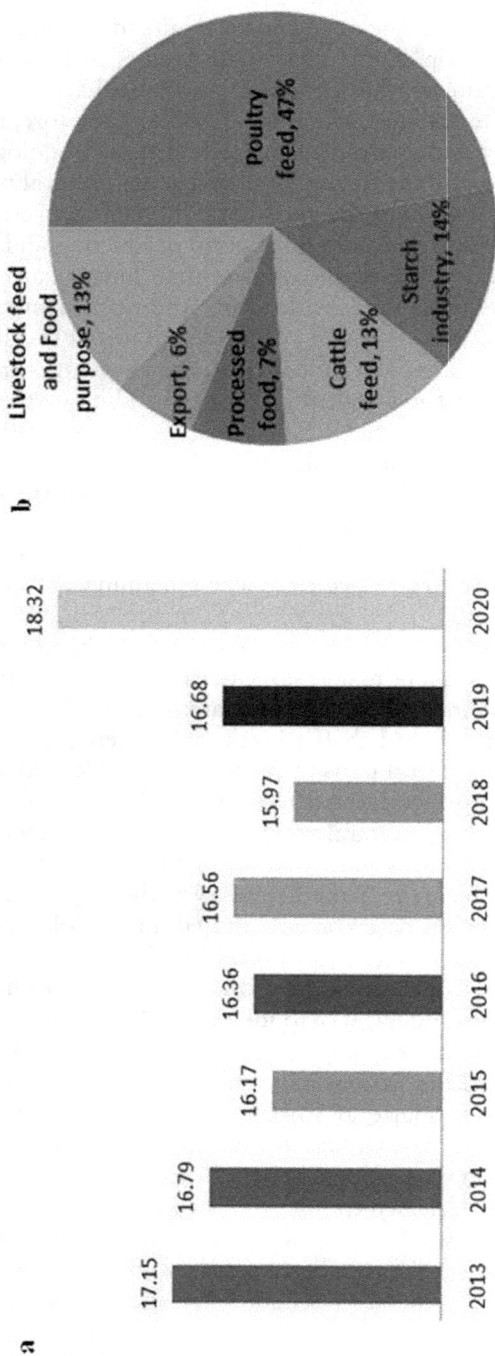

Figure 9.2 (a) Contribution of agriculture to the GDP for India from the year 2013 to 2020. (b) Distribution of utility of maize in different fields in India [2].

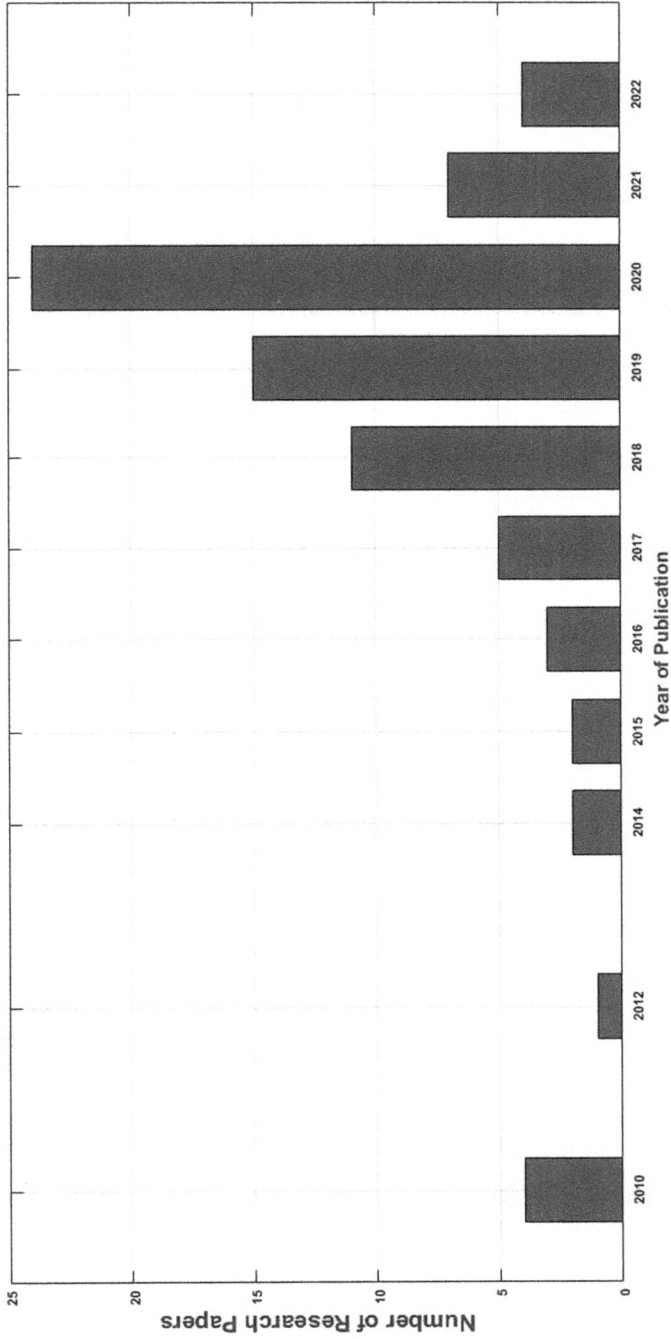

Figure 9.3 Number of research papers taken for this review work (year-wise).

9.2 CLASSIFICATION OF MAIZE DISEASES

The different types of diseases can infect maize plants at the different parts like stalk, leaf, ear, and tassel. These diseases are classified into four different categories based on their causes of illness [6], and the details are mentioned in Tables 9.1 to 9.4 and the respective images showing parts of plants infected by the different categories of diseases along with the respective healthy parts are shown in Figures 9.4 to 9.7.

Table 9.1 Various category of maize leaf diseases along with their symptoms

Name of the disease	Cause	Symptoms
Downy Mildews	Fungus	Initially, leaves and leaf sheaths show chlorotic striping and dwarfing. The leaves are thinner and more upright than healthy plants, with a white, feathery growth on both surfaces. As the plant matures, the leaves may acquire signs such as mottling and white striped leaves.
Curvularia Leaf Spot	Fungus	A tiny necrotic or chlorotic light-colored halo patch may appear on the affected leaf. When fully formed, the lesions are around 0.5 cm in diameter.
Maize Chlorotic Mottle Virus	Viruses and Mollicutes	Infected leaves have small chlorotic dots that coalesce into wide chlorotic stripes along the veins. These chlorotic streaks contrast with dark green tissue when seen in the light.
Gray Leaf Spot	Fungus	Here, the lesions developed with small and regular but elongated brown-gray necrotic spots, which appear and grow parallel to the veins.
Turcicum Leaf Blight	Fungus	Early symptoms emerge as tiny dots on somewhat oval, water-soaked leaves on lower leaves. An elongated, spindle-shaped necrotic lesion may form. It can grow in number and intensity as the plant matures, eventually causing leaf burn.
Bacterial Leaf Stripe	Bacteria	Leaves acquire numerous tiny pale-green lesions that might grow along veins, causing noticeable striping.
Yellow Leaf Blight	Fungus	Symptoms appear on maize's lower leaves. Initially, tiny, watery spots develop into elongated bands of gray-green to light brown lesions. Later spore-induced spots appear on top leaves.

(Continued)

Table 9.1 (Continued)

Name of the disease	Cause	Symptoms
Leptosphaeria Leaf Spot	Fungus	Here, the lesions are small, necrotic coalescing into concentric necrotic spots.
Maydis Leaf Blight	Fungus	Initially, the lesions are small and diamond-shaped. When they mature, they become elongated. The final lesion shape is rectangular and 2–3 cm long because the growth is limited by adjacent veins that may result in the complete burning of large areas of the leaves.
Stewart's Wilt	Bacteria	The lesions are initially water-soaked and have uneven borders that run the length of the leaves, which are pale yellow. The infection spreads to the stem in certain situations, resulting in plant stunting, withering, and death.
Maize Dwarf Mosaic Virus	Viruses and Mollicutes	Infected maize plants produce a mosaic of green color on the younger leaf bases. The mature plant's leaves turn purple or purple-red.
Polysora Rust	Fungus	Small size and light orange pustules and extra circular than the common rust. At the matured stage of the plant, the pustules turn dark brown.
Common Rust	Fungus	Symptoms include small and elongated powdery bumps that crop on either side of the leaves. The bumps are brownish in the beginning of infection, but the outer-most layer ruptures as the plant grows older, and the lesions turn black.
Sugarcane Mosaic Virus	Viruses and Mollicutes	Here, the infected maize plants are shorter in height, and the leaves show chlorosis, and it may die when the plant approaches flowering, which may cause no ear development.
Tropical Rust	Fungus	Pustules are small and found beneath the epidermis, which may vary in shape from round to oval. Sometimes, these bumps may look black-rimmed, but the center remains light. The lesion appears white to pale yellow at the center of the bump.
Zonate Leaf Spot	Fungus	The disease mainly occurs on older leaves in which the lesions are small and necrotic that produce large, concentric, and necrotic rings.

(Continued)

Table 9.1 (Continued)

Name of the disease	Cause	Symptoms
Banded Leaf And Sheath Blight	Fungus	The typical concentric or ring-like patches that cover significant areas of diseased maize leaves and husks may appear as symptoms.
Maize Fine Stripe Virus	Viruses and Mollicutes	Symptoms develop and can be observed after two weeks of infection. It starts with tiny, isolated chlorotic spots that can be seen by bringing leaves up to the light. Later, the spots become more numerous and fuse.
Tar Spot Complex	Fungus	The symptoms first appeared on the lower leaves before flowering, and the virus later extended to the younger leaves. It may seem like black, elevated, shiny areas when infected foliar tissue becomes necrotic and dies as necrotic tissue around the tar spot, causing the foliage to burn completely.
Phaeosphaeria Leaf Spot	Fungus	Small, pale green spots, which are spherical to somewhat elongate and become bleached, then necrotic, with a dark brown edge, may appear as disease symptoms.
Maize Streak Virus	Viruses and Mollicutes	A week after infection, the plant develops small as well as round and dispersed spots on the youngest leaves. The number of dots grows parallel to the leaf veins with plant growth. Mature leaves may have chlorosis with broken yellow streaks along the veins, contrasting with regular foliage's dark green.
Anthracnose Leaf Blight	Fungus	In the early and mature phases of plant development, the leaves become uneven with oval-to-elongated lesions with yellow-to-reddish-brown edges. The same can be seen in the upper part of the leaves of maize plants that have already developed stalk rot signs.

The maize plants can also get infected by physiological diseases due to different types of nutrient deficiency [7], which is given in Table 9.5 and their respective images are shown in Figure 9.8 for a better understanding of these diseases.

Table 9.2 Classification of maize stalk diseases along with their symptoms

Name of the disease	Cause	Symptoms
Charcoal Stalk Rot	Fungus	The abnormal drying of upper leaf tissue is the first sign, which appears only after flowering. Internally, vascular bundles shred with dark staining on stems. Here, the kernels are also turned completely black as the fungus attacked them.
Anthracnose Stalk Rot	Fungus	Infection symptoms can be noticeable starting from narrow as well as elongated dark lesions, which are initially brown and later turns black. As pith tissues are destroyed, premature wilting with shredded vascular bundles turning dark brown.
Maize Lethal Necrosis	Viruses and Mollicutes	The most unusual symptoms of this maize plant disease are dwarfing and striping along the veins.
Stenocarpella Stalk Rot	Fungus	This disease symptom develops where rotting has occurred as plenty of spore structures appear on the surface of internodes, which weaken the stalks and it can break easily during strong winds and rains.
Maize Bushy Stunt	Viruses and Mollicutes	The most typical symptom of this disease is mild chlorosis on young leaves, which progresses to purple-red tips as the plant matures.
Fusarium and Gibberella Stalk Rots	Fungus	Tiny and the lowest internodes produce dark-brown blemishes, while wilted plants remain standing when dry. But when the infected stalks are split, the phloem appears dark brown, and in the last stage of infection, the surrounding tissues become discolored while the pith is shredded.
Brown Spot	Fungus	Brown lesions on nodes and internodes. The leaves have chlorotic dots, whereas the mid-ribs have round dark brown patches. In severe infestations, they might combine and cause stalk rotting.
Botryodiplodia Stalk Rot	Fungus	Maize plants become dry along with shiny kernels, and husk leaves can also turn black and be fragmented.
Bacterial Stalk Rot	Bacteria	Infected plants have a dark color and water soaking at the stalk base and may die after tasseling. Bacterial decomposition can provide an unpleasant odor.
Corn Stunt	Viruses and Mollicutes	Leaf reddening or purpling and chlorotic lines at the base of younger leaves are two of the most visible symptoms. Foliar signs include stunted plants, bare ears at several nodes, and immoderate branching of root.
Pythium Stalk Rot	Fungus	Infected maize plants exhibit soft, wet, black basal internodes. Internodes often twist, although plants can thrive until all vascular bundles are compromised.

Table 9.3 Classification of maize ear diseases along with their symptoms

Name of the disease	Cause	Symptoms
Downy Mildew	Fungus	Infected maize plants cannot produce offspring or may have poor seed sets.
Botryodiplodia/Black Kernel Rot	Fungus	Ears affected by this disease have a deep black color with lustrous kernels, and husk leaves may become black and fractured.
Stewart's Wilt	Bacteria	The infection may spread to the stem, leading to general stunting, withering, and plant death.
Hormodendrum Ear Rot	Fungus	Dark brown or sometimes green streaks appear on both bases of cob as well as kernels. Damaged ears become light-weighted and black, causing physical injury to kernel tips.
Penicillium Ear Rots	Fungus	There is a pale blue and green powder in between the cob and that of kernels. If the fungus-infected grains usually grow, they will bleach and streak.
Fusarium and GibberellaEar Rot	Fungus	Infected kernels begin with white mycelium creeping down the tip, turning reddish pink. In some cases, this disease causes cottony growth on infected grains and white streaks on the pericarp. This disease's fungus creates mycotoxin, which is toxic to various animal species.
Charcoal Ear Rot	Fungus	Symptoms include light yellow kernels with black streaks below the pericarp and a loose, junky ear. The cob has easy-to-remove grains and little, spherical sclerotia on the surface.
Maize Rough Dwarf Virus	Viruses and Mollicutes	A month after birth, seedlings show symptoms. Infected plants appear crimson and develop no ear or nubbins with twisted tips.
Nigrospora Ear Rot	Fungus	It is hollow within; thus, it is light in weight. Close inspection of cob tissues reveals tiny black spore masses on kernel tips. The discolored kernels are easy to remove.
Maydis Leaf Blight "T" Strain	Fungus	This causes a rectangular or oval lesion with a 2–3 cm length. Large parts of leaves may be burned entirely.
Tar Spot Complex	Fungus	The diseased ears are light in weight and have detached kernels, with many germinating prematurely toward the ear's tip.
Maize Stripe Virus	Viruses and Mollicutes	Infected plants have stunted growth and twisted tassels. Normal ear development and yield are also reduced.

Head Smut	Fungus	Instead of the ear, the plant develops tasseling and silking stages with black and loose spore masses.
Aspergillus Ear Rots	Fungus	This disease produces black, powdery spore masses that coat kernels and cobs. It can be yellow-green spore masses or ivy green due to parasitic.
Ergot, Horse's Tooth	Fungus	Early signs of infection include sclerotia, pale, mushy, and slimy. When the maize plants approach for silk, the infected sclerotia germinate and grow multiple head-like structures that release new spores.
Common Smut	Fungus	Whitish galls replace individual kernels. They emit black masses of spores that infect maize plants the next season.
Gray Ear spot	Fungus	Small elongated brown-gray necrotic patches appeared and grew parallel to the veins.
Stenocarpella Ear Rot	Fungus	This disease symptom appears on the surface of internodes where rotting has occurred, weakening the stalks and causing them to break readily during heavy winds and rains.

Table 9.4 Classification of maize tassel diseases along with their symptoms

Name of the disease	Cause	Symptoms
Head Smut	Fungus	Symptoms include aberrant tassel development, spore masses inside male florets, and exposed vascular bundles. But symptoms appear only after tasseling and silking.
Downy Mildews	Fungus	The leaves are thinner and more upright than healthy leaves, which may have a white, downy growth on both sides. The leaves may develop mottling and white stripes as the plant ages. Lesser leaves and leaf sheaths show dwarfing and chlorotic striping.
Corn Stunt/Maize Bushy Stunt	Viruses and Mollicutes	The most typical symptom of this disease is mild chlorosis on young leaves, which progresses to purple-red tips as the plant matures.
Common Smut	Fungus	Prominently closed white galls replace individual kernels. When the galls decompose, they release black spore masses that ultimately infect maize plants the following season.
Maize Rough Dwarf Virus	Viruses and Mollicutes	The tassels and upper leaves are deformed. Younger leaves curl upwards, with characteristic vein overgrowth at the base as they mature. Symptoms appear after one month in seedlings. As the disease spreads, plants turn reddish and produce no ear or only nubbins, which are occasionally twisted at the tip.
Maize Stripe Virus	Viruses and Mollicutes	Tiny chlorotic dots on the leaves appear first, then narrow parallel chlorotic stripes along with the younger leaves. The infection causes tassels to stutter and bend. Ear development and yield are also diminished.
Maize Chlorotic Mottle Virus	Viruses and Mollicutes	Infected leaves have minute chlorotic dots that coalesce into wide chlorotic stripes along the veins. These chlorotic streaks contrast with dark green tissue when illuminated.
False Head Smut	Fungus	The only symptoms are dark-green spore masses in the tassels of a few isolated male florets. Normal smut does not produce galls; however, fake head smut does.
Bacterial Leaf Stripe	Bacteria	Leaves acquire a plethora of microscopic pale-green lesions that can grow along veins, resulting in visible striping in some cases.

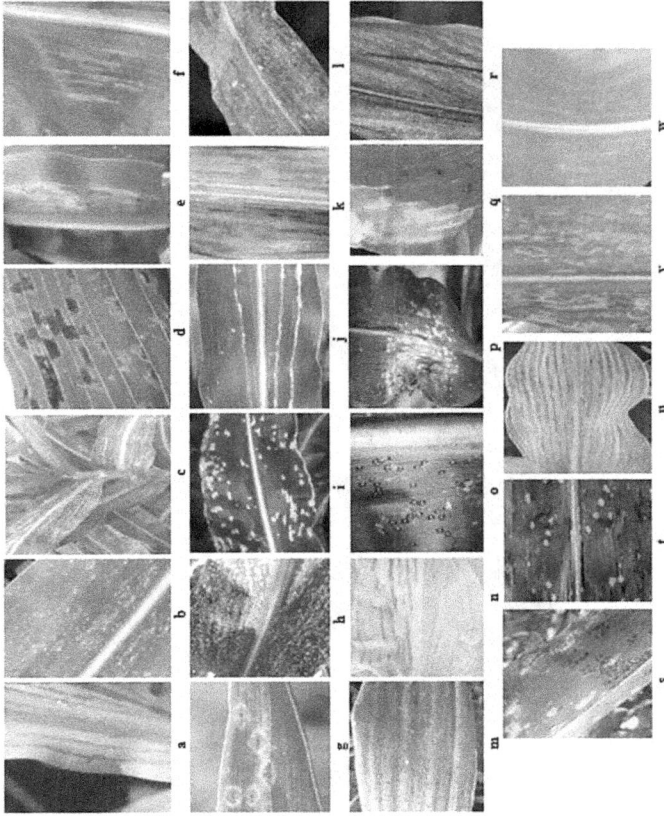

Figure 9.4 Different kinds of maize leaf images. (a) Leaf infected with Downy Mildews. (b) Leaf infected with Curvularia leaf spot. (c) Leaf infected with Maize chlorotic mottle virus. (d) Leaf infected with Gray leaf spot. (e) Leaf infected with Turcicum leaf blight. (f) Leaf infected with bacterial leaf stripe. (g) Leaf infected with yellow leaf blight. (h) Leaf infected with Leptosphaeria leaf spot. (i) Leaf infected with maydis leaf blight. (j) Leaf infected with Stewart's wilt. (k) Leaf infected with Maize dwarf mosaic virus. (l) Leaf infected with Polysora rust. (m) Leaf infected with common rust. (n) Leaf infected with Sugarcane mosaic virus. (o) Leaf infected with Tropical rust. (p) Leaf infected with Zonate leaf spot. (q) Leaf infected with Banded leaf and sheath blight. (r) Leaf infected with Maize fine stripe virus. (s) Leaf Infected with Tar spot complex. (t) Leaf infected with Phaeosphaeria leaf spot. (u) Leaf infected with Maize streak virus. (v) Leaf infected with Anthracnose leaf blight. (w) Healthy leaf.

Figure 9.5 Different kinds of maize stalk images. (a) Stalk infected with Charcoal rot. (b) Stalk infected with Anthracnose stalk rot. (c) Stalk infected with Maize lethal necrosis. (d) Stalk infected with Stenocarpella stalk rot. (e) Stalk infected with Maize bushy stunt. (f) Stalk infected with Fusarium and gibberella stalk rots. (g) Stalk infected with Brown spots. (h) Stalk infected with Botryodiplodia stalk rot. (i) Stalk infected with Bacterial stalk rot. (j) Stalk infected with Corn stunt. (k) Stalk infected with Pythium stalk rot. (l) Healthy stalk.

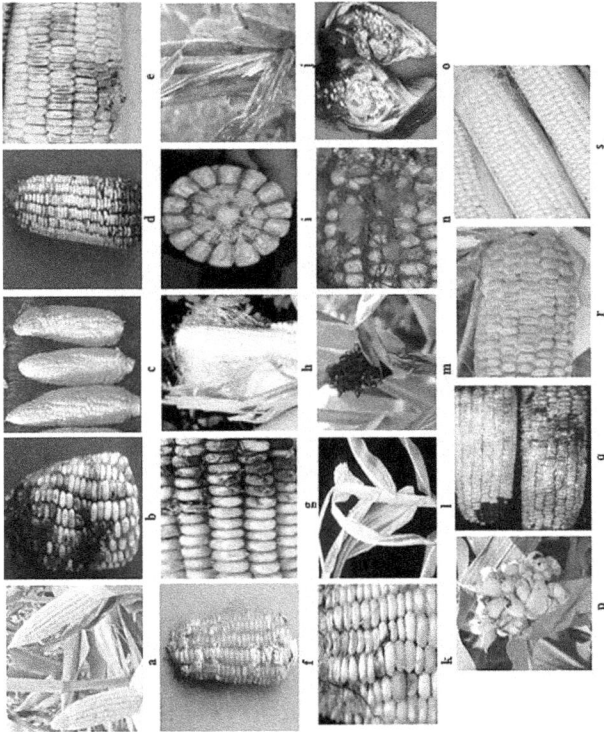

Figure 9.6 Different kinds of maize ear images. (a) Ear infected with Downy mildews/Corn Stunt. (b) Ear infected with Botryodiplodia/Black kernel rot. (c) Ear Infected with Stewart's wilt (d) Ear infected with Hormodendrum ear rot. (e) Ear Infected with Penicillium ear rots. (f) Ear infected with Gibberella ear rots. (g) Ear Infected with Charcoal ear rot. (h) Ear infected with Maize rough dwarf virus. (i) Ear Infected with Nigrospora ear rot. (j) Ear infected with maydis leaf blight "T" strain. (k) Ear Infected with Tar spot complex. (l) Ear infected with Maize stripe virus. (m) Ear infected with Head smut. (n) Ear infected with Aspergillus ear rot. (o) Ear infected with Ergot. (p) Ear infected with Common smut. (q) Ear infected with gray ear rot. (r) Ear infected with Stenocarpella ear rot. (s) Healthy Ear.

Figure 9.7 Different kinds of maize tassel images. (a) Ear infected with Head smut. (b) Ear infected with Downy mildews. (c) Ear infected with Maize bushy stunt. (d) Ear infected with Common smut. (e) Ear infected with Maize rough dwarf virus. (f) Ear infected with Maize stripe virus. (g) Ear infected with Maize chlorotic mottle virus. (h) Ear infected with False head smut. (i) Healthy tassel.

Table 9.5 Different types of physiological maize diseases along with their symptoms

Name of the disease	Symptoms
Nitrogen deficiency	When maize plants lack nitrogen, they turn pale green and grow a yellow "V" shape on their leaves from the leaf end to the leaf collar.
Phosphorous deficiency	Symptoms of this nutritional deficit include reddish-purple lower leaves and dark green plants.
Potassium deficiency	Here, the symptoms develop from lower leaves to upper leaves, and the leaf margins may turn yellow and brown.
Sulfur deficiency	Younger leaves have yellow color striping.
Zinc deficiency	The signs of zinc deficiency first develop in the leaves' center and then spread outward. When the plant matures, the upper leaves grow broader bands of yellow hue that turn pale brown or gray necrosis like dead spots.

9.3 IDENTIFICATION OF MAIZE DISEASES USING COMPUTER INTELLIGENCE

Food security depends upon plant diseases because agricultural food production will decrease when plant diseases are not being discovered at an early stage or at the proper time. To make efficient management and decision-making systems for agriculture to increase productivity, detection of plant diseases at an early stage is becoming more imperative. In most situations, the primary source for identifying plant illnesses is the leaves of the plants,

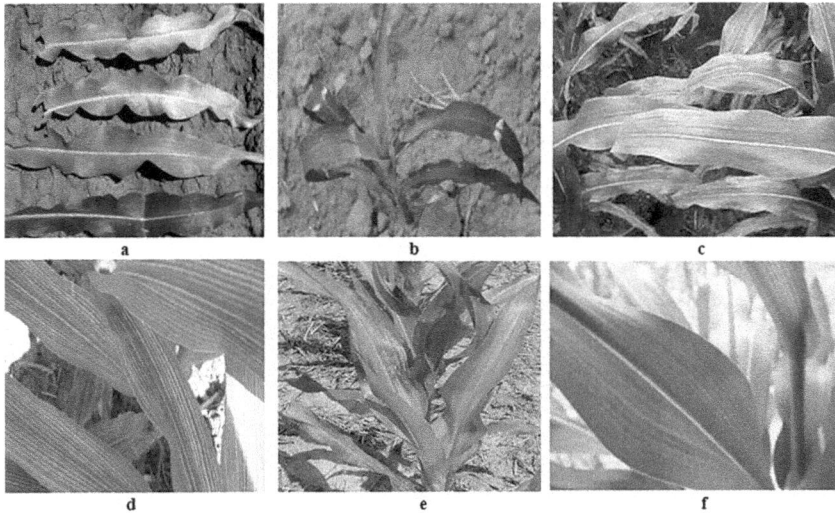

Figure 9.8 Nutrition-deficient maize plant images. (a) Nitrogen deficiency. (b) Phosphorous deficiency. (c) Potassium deficiency. (d) Sulfur deficiency. (e) Zinc deficiency. (f) Healthy image.

as the manifestation of the diseases might begin to emerge on the leaves. Moreover, the disease-infected plants have some unique marks, patterns, or wounds on leaves, flowers, fruits, and stems. Generally, it is observed that the disease identification in maize plants and also for other plants is made by the agricultural expert that is subjective and time-consuming or by the farmer itself based on their experience, which is laborious and inefficient because farmers may misjudge and can use some inappropriate drugs, which can decrease their productivity and may lead to financial loss. These diseases can be recognized and classified by using image processing techniques easily and efficiently.

This section presents the basic methodologies based on computational intelligence and image processing techniques. Figure 9.9 shows a traditional process diagram for the disease detection for maize plants and other plants, whereas a generalized block diagram for classifying different types of maize and other plant diseases using ML and DL classifiers is shown in Figure 9.10.

9.3.1 Image acquisition

The initial stage in every image processing system is image acquisition. It's a process of converting captured photographs to the proper output format so that they may be processed further for future use. Any image acquisition aims to convert an optical image or real-world data into a numerical array that can then be manipulated or edited on a computer. In any study, the dataset is crucial. So, before beginning any type of research, the first and

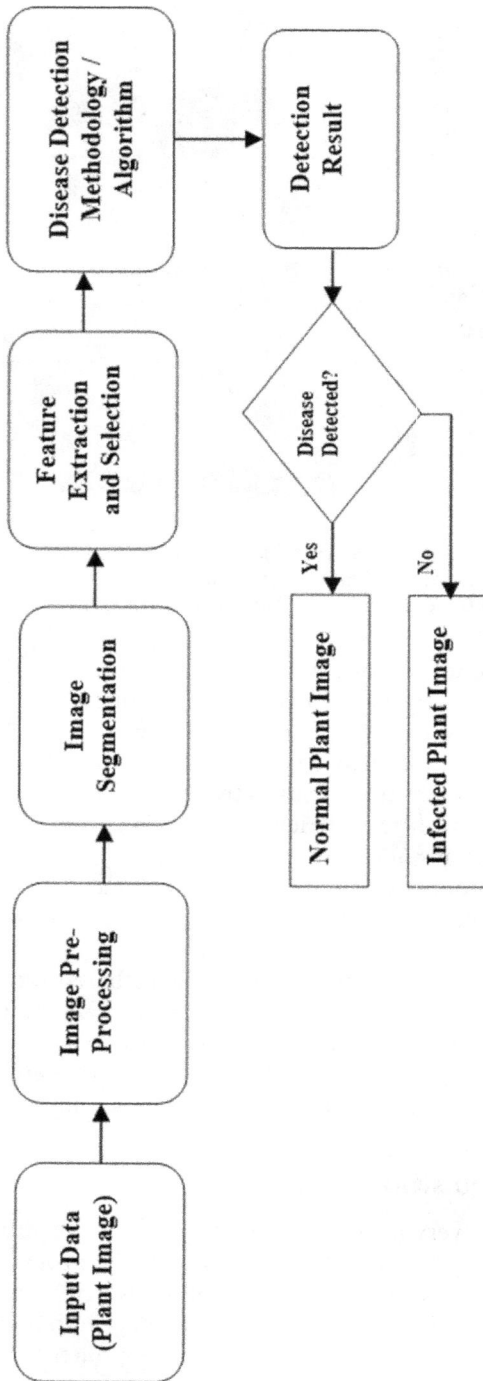

Figure 9.9 Generalize process diagram for plant disease detection.

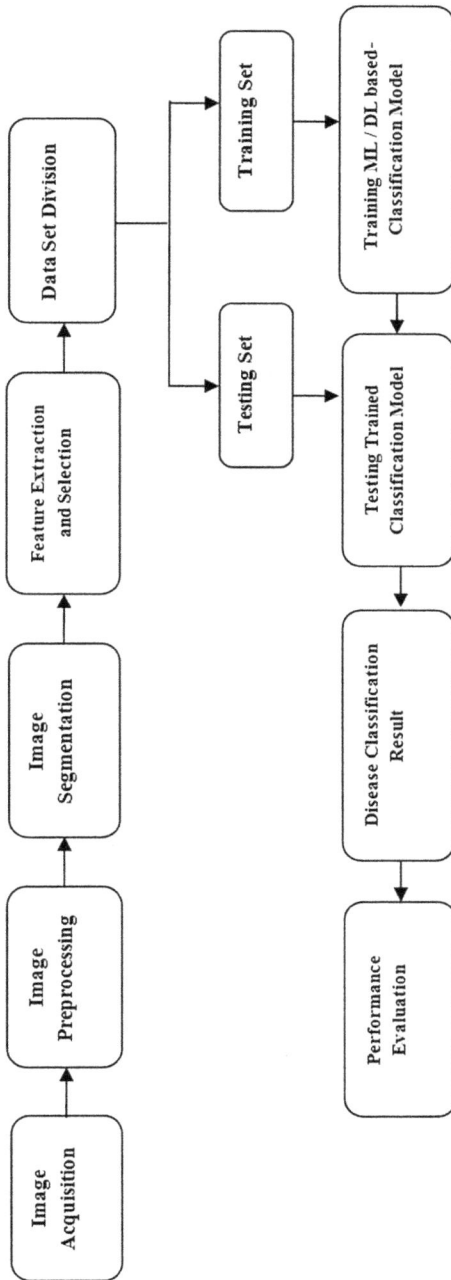

Figure 9.10 Generalize block diagram to entail the classification of plant diseases by using ML and DL classifier.

most important task is to collect and process as much data as possible. Only a sufficient amount of data can provide the highest level of accuracy and make any research effective. The maize picture dataset can be created in one of two ways, or both methods can be followed: 1. Manually capture and construct a dataset for infected and healthy maize plant photos using any good quality camera from any corn harvesting farm. 2. Each type of maize photograph can be searched and collected using web search engines such as Google, or the maize image dataset can be acquired from dataset websites such as Plant Village [8], PlantDoc [9], Kaggle [10], OSF storage [11], and so on.

9.3.2 Image preprocessing

The goal of preprocessing is to enhance the image's quality to analyze it more effectively. We can reduce unwanted distortions and boost some essential properties for the application we are working on by preprocessing; as a result, we can highlight the region of interest, that is, the disease-infected area in maize images. The image preprocessing functionalities may differ depending on the application. Some of the image preprocessing techniques [12, 13] are Pixel brightness transformations/Brightness corrections, Geometric Transformations, Affine Transformation, Perspective Trans-formation, Image Filtering, Fourier transformation, and Data Augmentation [14].

9.3.3 Image segmentation

In digital image processing, image segmentation divides a picture into meaningful portions or zones, frequently depending on the image's pixel attributes. Image segmentation techniques include threshold, Area, edge, watershed, and clustering. Color or shape similarity can be used to cluster pixels in an image.

9.3.4 Feature extraction and selection

In computer vision and machine learning, feature vectors are used to describe objects in an image [15]. Using feature extraction method, the captured object is sorted into its applicable class. The structures' forms, sizes, origins, colors, and textures were used to describe the diseased part in the image. Agricultural feature extraction approaches include form, color, and texture, each with its challenges. These techniques are employed in order to identify diseases on the basis of texture, size, and color. On the other hand, color feature is vital because each disease has its unique color. Texture and shape can also assist detect diseased plants. Compared to prior techniques, deep learning offers the advantage of automatic feature extraction, which increases accuracy.

9.3.5 ML/DL classification models

Image classification is a supervised learning issue where a model is trained to recognize target classes or objects in images. Initially, computer vision models employed raw pixel data. The last phase uses a classification model to identify the leaf disease. The model should be trained using known pathologies and learning methods. Automated learning algorithms can learn from input data to achieve a specific goal. Its powerful computer offers new agricultural possibilities. Some ML classifiers are Naive Bayes (NB), Decision Trees (DT), K-Nearest Neighbors (KNNs), Support Vector Ma-chines (SVMs), Random Forests (RF), Recurrent Neural Networks (RNNs), and Extreme Learning Machines (ELMs), whereas deep neural networks (DNNs) are used to develop hierarchical data representations at multiple degrees of abstraction [16]. Convolutional neural networks (CNNs) are a strong and primary deep learning tool for modeling complex processes. These include GoogleNet, VGGNet, ResNet, AlexNet, MobileNet, LeNet, RCNN, DenseNet, ARNet, and PARNet.

9.3.6 Performance evaluation

Performance measurements are used to evaluate algorithms. These indicators include accuracy, sensitivity, precision rate, Area under the curve,and so on [15]. The accuracy distinguishes between accurate and erroneous samples. Sensitivity is the rate of accurately identified pictures. Area under the curve indicates how a classifier evaluates a positive or negative instance. This method also uses F1 score, informed-ness, G-measures, and likelihood ratio. Time is vital in real-time circumstances; hence, many researchers calculate time to verify their ideas.

9.4 RECENT RESEARCH WORKS ABOUT THE IDENTIFICATION AND CLASSIFICATION OF MAIZE DISEASES

Many researchers have contributed to detecting and classifying maize plant diseases through their studies. This section provides a detailed assessment of some of the effective research work done in the last 10 years to detect, identify, and classify maize plant diseases by utilizing image processing approaches based on computational intelligence methodologies. At the end of this section, the detailed summary about the related research work on above said objective is presented in Table 9.6. Ishengoma et al. [17] created a hybrid model by integrating VGG16 and InceptionV3 in parallel to detect Fall Armyworms infested maize leaves, since VGG16 and Incep-tionV3 provide better accuracy than XceptionNet and ResNet50 models. The input photos were applied to both models at the same time. For classification, they

Table 9.6 Summary of research work for the identification and classification of afflicted maize plant

Reference	Methodology/ model used	Data set source	Sample size	Imagesize	Data augmentation	Accuracy
[17], 2022	VGG-16, Inception-V3	Manually collected by using Quad-copter drone at different farms in East Africa.	10,000	150 × 150	Yes	96.98%
[18], 2022	SKPSNet-50	Manually Captured by using CanonEOS 80D digital camera at different locations of China.	2904	448 × 448	Yes	92.2%
[19], 2021	VGG16,VGG19, MobileNetV2, InceptionV3	Manually collected by using Phantom 4 Pro V2 (Quad-copter drone)	11,280	224 × 224	Yes	99.92% (VGG16) 99.67% (VGG19) 100% (MobileNetV2) 100% (InceptionV3)
[20], 2021	AlexNet (TCI-ALEXN)	Plant Village Database and Experimental field of Jilin Agricultural University	3852	256 × 256	Yes	93.28%
[21], 2020	DenseNet	Manually gathered from different Sources	12,332	250 × 250	Yes	98.06%
[22], 2020	SVM, NB, KNN, DT, RF	Unknown	3823	100 × 100	No	77.56% (SVM) 77.46% (NB) 76.16% (KNN) 74.35% (DT) 79.23% (RF)
[23], 2020	Deep CNN	Plant Village Database and manually collected from corn plantation	4382	150 × 150	No	88.46%
[24], 2020	CNN	Plant Village Database	4354	256 × 256	Yes	97.6%
[25], 2019	Modified LeNET	Plant Village Database	3852	64 × 64	No	97.89%
[26], 2019	KNN, SVM	Collected by using Samsung Digital camera PL200	200	Unknown	No	85% (KNN) 88% (SVM)

Reference, Year	Method	Data source	Number of images	Image size	Augmentation	Accuracy
[27], 2019	CNN	Captured using a Google Pixel 3 Smartphone camera	100	10 × 20 × 3 (height × width × RGB)	No	92.85%
[28], 2018	SVM	Plant Village Database	2000	Unknown	No	83.7%
[29], 2018	Multi-channel CNN	collected in the fields, taken with a Canon EOS700D	10,820	5,312 × 2,988	Yes	92.31%
[30], 2017	GoogLeNet and Cifar10	Plant Village Database and downloaded from Google.com	3060	224 × 224	Yes	98.90% (GoogLeNet) 98.80% (Cifar10)
[31], 2017	DMS-Robust AlexNet	Downloaded from different sources namely: Global AI Database, Plant Village and Google search engine.	12,227	256 × 256	Yes	98.62%
[32], 2017	CNN	Captured by using SONY a600 and Canon EOS Rebel at Musgrave Research Farm, Aurora	1769	224 × 224	No	96.7%
[33], 2017	CNN	Plant Village Database	8506	256 × 256	Yes	97.09%
[34], 2015	KNN	Unknown	100	32 × 32	No	90%
[35], 2014	SVM	Unknown	200	Unknown	No	83.2%

forwarded the combined outputs generated by both models to the fully connected layers (using the SoftMax activation layer). Their proposed model was dubbed the VGG16-InceptV3 model. To photograph the maize plants, the author used UAV (Unmanned Aerial Vehicle) technology. They employed TensorFlow 2.0 and Python programming to develop the proposed CNN model using 10,000 maize leaf images for the experimental study. The proposed hybrid model attained a high level of accuracy of 96.98% in their experiments. The key advantages of this model are that it takes much less time to train than previous models, and farmers are automatically notified of the results by SMS and email once the classification process is complete. The proposed VGG16-InceptV3 hybrid model has to be enhanced so that farmers can apply pesticides just to the infected areas at the beginning of the infection, reducing treatment time and expense. This can be performed by locating Fall Armyworm-infected maize plants in their exact location. Zeng et al. [18] introduced the SKPSNet-50 CNN model to recognize maize leaves afflicted by Downy mildews, Rust, Leaf blight, Corn borer as well as Locust eating. They created the proposed model by replacing the RELU with the 'Select Kernel–Point–Swish B' activation function in the base network Res-Net-50, and they also used the 'Swish B' to enhance the feature extraction trait of diseased leaves having tiny spots and uneven shapes. The author used Python 3.7.3 with the PyTorch 1.3.1 framework as a software environment and manually captured around 2904 maize images from both infected and healthy leaves using a Canon EOS 80D digital camera in various locations of China. The results reveal that the proposed model correctly distinguishes between infected and healthy maize leaves with a 92.2% accuracy. The proposed model features numerous parameters and a significant calculation amount as compared to various light-weight networks. Despite having fewer parameters, it has a higher F1 score than Inception-v3 and Inception-v4. The F1-score of the suggested model is just 0.001 lower than that of inception-ResNet-v2.

Ishengoma et al. [19] used various CNN models like VGG16, VGG19, Mo-bileNetV2, and InceptionV3 to automatically detect maize leaf images infected by Fall Armyworms. They have used remote sensing technology named UAV, which stands for Unmanned Aerial Vehicle to capture 1089 numbers of maize leaf images from the field and utilized TensorFlow 2.0 and python programming language to build the CNN models. For the experiment, the author used various image augmentations techniques like mirroring, rotating, shifting, and photometric transformation for expanding the captured images to a larger image dataset of 11,280 numbers to train and test the designed CNNs. In this proposed approach, the models are simulated by both the original and the modified images (Shi-Tomas corner detection techniques do image modification). Their results show that the models VGG16, VGG19, Mo-bileNetV2, and InceptionV3 successfully detect the defected maize leaves out of the healthy leaves at 99.92%, 99.67%, 100%, and 100% when trained with the modified images. Xu et al. [20] propound a global pooling method based on a multi-scale neural network to detect maize leaf diseases. In this method, the ability of AlexNet feature extraction is enhanced by adding a module of convolution layer long with a new Inception layer, where the Inception layer is combined to the ConCat layer. The authors used a globally pooled layer in place of a fully connected layer and transferred learning to find the solution regarding the overfitting problem in AlexNet for inadequate data sample. The experiment used 3852 and 907 numbers of color leaf images for three different disease categories and that of healthy leaf, which are collected from the Plant Village database and experimental arena of Jilin Agriculture University, respectively. The obtained result shows that their method achieved an accuracy of 93.28%, which is 2.42% improved than the traditional AlexNet model (90.86%).

Waheed et al. [21] optimized a dense CNN for the purpose of identifying and classifying maize leaf diseases such as gray leaf spot, northern leaf blight, and common rust. They constructed the classification model using a deep CNN. They begin by splitting the complete data set into two categories, training and testing, and then normalize all of the image's pixel values by separating into parts each pixel by 255.0 to arrange them within the range of 0 to 1. The authors used rescale, rotation, width shift, height shift shear, and zoom procedures with 1/255.0, 45, 0.2, 0.2, 0.4, and 0.3, respectively. The modified photos were then provided as inputs to the CNN model. Due to the fact that training DL models can take several hours, days, and weeks and one can lose a lot of effort if the run is abruptly terminated, the authors stored the weights of the most efficient model in a file named as weights.h5 can be utilized for training or other reasons later. They employed 12,332 photos of healthy and sick maize leaves in the experiment, and the results indicate that the model correctly identified the above diseases 98.06% of the time. Panigrahi et al. [22] have used several machine learning techniques for maize diseases identification. The classifiers used here are KNN, NB, DT, RF, and SVM for finding out various plant diseases of maize leaves. The architecture of their proposed method consists of several components like image acquisition, image pre-processing, image segmentation, feature extraction, and classification. First, the collected maize leaf RGB images are converted to grayscale images to provide accurate results. Then, the images are divided into several objects or regions using two different image segmentation methods based on similarities and discontinuities to extract useful information. Then, the features are extracted from the images and categorized on the basis of shape, color, and texture. Then, all this extracted information is used to train the above-said models. The authors used 3823 numbers of maize images (healthy leaf images-1162, common rust images-1192, gray leaf spot-513, and northern leaf blight-956) collected from the Plant Village database to investigate the performance of the supervised models on the basis of accuracy. Their experimental result shows that different used classifiers like SVM, NB, KNN, DT, and RF give 77.56%, 77.46%, 76.16%, 74.35%, and 79.23% of classification accuracy, respectively. Mishra et al. [23] have achieved an average identification accuracy of 88.46% by improving the performance of the deep neural network by modulating the hyper parameters and regulating the pooling combinations on a system with GPU. The deep CNN is developed by arranging various layers, namely, convolutional Layers (to extract the features from input images) where RELU is used as an activation function in each layer except at the output layer, max-pooling layer (to aggregate the statistics of feature map), drop out layer (to improvise capabilities of the trained model), flatten layer (to transform the feature matrix into a vector), and dense layer (to execute the linear operation on input, where the input is multiplied with weight matrix); finally, after addition of a bias vector with the output of dense layer, the output is produced. In the above proposed method, 4382 numbers of maize leaf images belonging to three categories (healthy leaf images-1447, common rust images-1558, and northern leaf blight-1377) are used to test the performance in terms of accuracy. To solve the problem of long convergence time and the small number of samples, Hu et al. [24] introduced a CNN recognition model based on data enhancement and migration learning. The proposed idea is based on the GoogLeNet model, which contains 22 layers of network structure characterized by the depth of the network. They made two improvements based on traditional CNN. The first one was to improvise robustness and generalization of the model by the data enhancement methods. The second one was to accelerate the process of training the network and decrease the degree of overfitting of the model by using migration. The improvement of the model, which was more suitable for their data output, was mainly to adjust the parameters of the full connection layer of the GoogLeNet model to reduce and refine the Softmax layer. In the experimental analysis, they have taken a total of 4354 numbers of healthy as well as infected Corn images (which are downloaded from the plant village dataset), and the investigation shows that the proposed model has achieved 97.6% of average accuracy for classification. The presented model successfully avoids the problems in traditional neural network classification, which makes it easier to fall into the local optimal solution and over-fitting due to the small amount of data.

Priyadharshini et al. [25] suggested a deep CNN-based architecture for classifying three distinct maize leaf diseases, using the LeNet deep network for classification. To begin, the picture data collection is preprocessed using Principal Component Analysis, which is known as dimensional-reduction algorithm to reduce the correlation between the features and to accelerate the feature-learning algorithm. The collected dataset of maize leaf images is then arbitrarily partitioned into two subsets: Training and Testing subsets. The proposed method trains the CNN in four distinct sections: forward pass (in which the CNN considers the training images and passes them through the entire network), loss function (most commonly defined as Mean Square Error), backward pass (in which the CNN determines which weights contribute to the loss and identifies a method to modify them to minimize the loss), and weight update (to update the weights of the filters in order to change the direction of the gradient). The authors trained and evaluated the proposed model using a total of 3852 maize picture datasets obtained from the Plant Village database. Their testing results indicate that the above model can accurately classify photos into several categories such as healthy, gray leaf spot, northern leaf blight, and common rust. Deshpande et al. [26] proposed an automated detection model using first-order histogram and Haar wavelet feature based on GLCM features to detect the fungal diseases in maize leaves. At first, the authors manually captured 200 numbers of images belonging to four different categories, namely, 50 numbers of healthy, 50 digits of northern leaf blight, 50 numbers common rust, and 50 numbers of leaf images with both diseases from agricultural domains at the Agricultural University, Dharwad. Then, all the images are preprocessed to correct the non-uniform illuminations by converting the RGB images into L * A * B color space to extract the luminance component (L), the Contrast-limited adaptive histogram equalization (CLAHE) is applied to the L component, and concatenated with A and B components. The image is converted from L * A * B color space to RGB space. Then, R, G, and B components of images are distinguished using the First-order Histogram features; 18 features (six features per component) per image are extracted. Features like horizontal, vertical, and diagonal coefficients are extracted using Haar wavelet, and features like contrast, correlation, energy, and homogeneity are extracted using GLCM (Gray-Level Co-occurrence Matrix). Finally, two classifiers, namely, KNN and SVM, are used to classify images on the basis of features extracted from the previous stages. The investigation shows that the above model classifies the diseases with 84% accuracy with the KNN (K = 5) classifier and 85% with the SVM classifier when using the RBF kernel. Still, the model shows the highest accuracy of 85% with the KNN (K = 7) classifier and 88% with the SVM classifier. Sibiya et al. [27] created a strategy for recognizing and classifying maize disease photos recorded with mobile phones using CNN. The proposed model incorporates a CNN with 50 hidden layers for the purpose of recognizing and classifying maize leaf illnesses from healthy leaves. Back propagation was used to train the network, and stochastic gradient descent (SGD) was utilized to learn the CNN model's parameters. The proposed model's average recognition precision was 92.85% in their experimental analysis, which included 100 images for each category (healthy leaf, gray leaf spot, northern leaf blight, and common rust). One of the advantages of this model is that it can be used by anyone who is not familiar with a high-level programming language, as the proposed model is built on a graphical user interface platform. Nonetheless, the proposed method's primary shortcoming was its limited sample size.

Aravind et al. [28] established a methodology for extracting textural aspects of maize leaf disease using the gray level co-occurrence matrix. They constructed the bag of features by extracting both SURF (where for all the training images the key-points were selected and the feature descriptors were evaluated by using the wavelet technique) and textural features (using histogram and gray-level co-occurrence matrix) for each image. Then, all of these extracted properties were fed to the classification algorithm, which was used to identify and classify sample leaves into distinct illness groups. Then, using a multi-class SVM, maize illnesses are categorized. The author evaluated the suggested model's performance using 2000 photos of maize leaves classified as leaf spot, rust, leaf blight, and healthy leaves obtained from the Plant village database. Due to the fact that the SVM classifier is adequate for smaller samples and is inadequate for large samples in terms of achieving high identification accuracy. Lin et al. [29] used multi-channel CNN models to alter the ratio of training to testing data, hence increasing the accuracy of diagnosing maize illnesses. The suggested model is composed of 13 major hidden layers and some auxiliary layers are put above these 13 levels, including seven RELU layers, two normalization layers, and two dropout layers, to limit overfitting and gradient diffusion. They created a standard database using homo-morphic filtering and ROI segmentation images for training and testing the suggested model. Following that, an enhanced database was created to boost the quantity and diversity of photographs. They conducted an experimental study on 10,820 photos comprising five prevalent diseases, including leaf blight, sooty blotch, brown spot, rust, and purple leaf sheaf. The model attained an average accuracy of 92.31%. Despite the fact that the above strategy produces positive results, the growing sample size and the respective extended time for training convergence have a detrimental effect on recognition accuracy. To increase the accuracy of diagnosing disease in the leaf of maize plant, it is critical to create a recognition prototype with a small sample and high identification accuracy.

Zhang et al. [30] suggested two enhanced deep learning models (GoogLeNet and Cifar10) for recognizing various leaf diseases in maize plant by tweaking model parameters, altering pooling combinations, incorporating the dropout operation, and adding the RELU function. To begin, they gathered 500 images from various sources and categorized them into numerous diseases. Then, they preprocessed the gathered photos and employed several augmentation approaches to increase the dataset from 500 to 3060 images (since 500 images are insufficient for efficient feature extraction and classification). The experimental result indicates that Google Net's average accuracy increases to 98.9% when combined with Cifar10, and to 98.8% when combined with Cifar10. Lv et al. [31] presented a maize leaf disease recognition prototype based on feature enhancement and Alexnet architecture (the backbone of this model). They are named DMS-Robust Alexnet, where dilated-convolution along with multi-scale convolution is integrated to improve feature ex-traction capability. They have also performed batch normalization to overcome network over-fitting along with improving the model's robustness while achieving improved convergence as well as accuracy by using the PRELU (instead of Traditional function RELU or sigmoid) and Adabound optimizer. In their proposed model, the data set is created by collecting maize disease images belonging to nine different categories, namely, rust, leaf spot, leaf blight, zinc deficiency, round spot, fall armyworm, and healthy from different sources mentioned in [8, 9]. Second, different data augmentation methods are applied to the collected dataset for minimizing the uneven distribution of samples and to procure disease feature enhancement images; all the data are enhanced by using the proposed Wavelet transform Donoho threshold and improved Retinex (WT-DIR). Third, the resultant enhanced data set trains the constructed DMS-Robust Alexnet. And finally, the trained model is used to identify the maize diseases. The experimental analysis used 12,227 numbers of images, and the results show that the presented method outperforms with 98.62% recognization accuracy. Dechant et al. [32] presented a method for identifying maize leaves afflicted with northern leaf blight that utilizes a CNN computational pipeline. The mentioned approach overcomes the limitations of finite data and the non-uniformity in the field-grown plant photos. They demonstrated that the proposed technique achieved 96.7% on test images. In the proposed scheme, the infected leaf was detected in three stages: stage 1 involved training several CNN to detect infected patches on the images; stage 2 involved using the trained CNN to generate heat maps depicting the occurrence of infection in each area of the images; and stage 3 involved using the generated heat maps to classify the entire set of images. The experimental study included a total of 1796 pictures, of which 1028 were of infected maize leaves and 768 were of healthy maize leaves. Additionally, the proposed technique attained a precision of 96.7%. Nonetheless, one of the primary disadvantages is that manually classified images are required to train the CNN, which takes some time. The advantage is that this method does not require specialist gear; consequently, the network and the proficiency it represents may be distributed across multiple users. Ouppaphan [33] demonstrated a lightweight CNN architecture trained to identify the corn leaves into one of four maize leaf disease categories. They trained the suggested CNN model using five-fold cross-validation from scratch in this approach. They employed three subsets of data in each fold: one for data augmentation, another for training, and the last subset for validation. The concept was to use the model that performed the best on the validation data to test the remaining subset of data. The method was performed five times to ensure that each data point was tested at least once. They used 8506 pictures of maize leaves divided into four distinct groups for the experimental study, and the inquiry revealed that the propound model attained an exactness of 97.09%. Given the lesser complexity of the aforementioned architecture, making it more apt for real-time inference on resource-constrained devices such as Smartphone, the achieved average accuracy is enough for the categorization of afflicted maize leaf.

Zhang et al. [34] used a KNN classifier to determine the afflicted maize plant leaf by using the extracted features from the segmented spot. At first, they preprocessed the collected maize leaf images (converting RGB to HSV format space, masking the green pixels, and removing the stem and the masked green pixels) before segmenting the spot components to get the recognizable disease spot (i.e., the disease feature vector). They have also applied histogram equilibrium to resize the images. And finally, they use all these processed images for training and testing the model. For the experiment, they used 100 numbers of maize leaf images belonging to five different categories, and the result shows that the recognition accuracy of their proposed model reached over 90%. But the proposed model has some limitations like the model remains prone to environmental effect and leads to over-segmentation, which may have an adverse effect on the result. Based on SVM, Zhang et al. [35] put forward a model to identify the afflicted maize leaf using the mobile terminal as one of the fundamental tools for collecting the disease information. The feature of this method is as identification results return to the terminal itself, it can improve the efficiency of agricultural producers. To provide an easy interface to the user for any modification in the SVM parameters and realize Remote diagnosis of disease, they have used C # and.NET to create web services and recognition programs for disease and publishing website connected to the database server, where anyone can take a picture of a maize leaf and upload it to the server or download image data of maize diseases from the designed server database. To investigate the performance of their proposed method, they used 200 images for five categories of selected diseases, namely, Northern and Southern leaf blight, corn rust, maize head smut, maize smut, and the investigation shows that the proposed method has achieved 83.2%average recognition accuracy.

9.5 RELATED WORK ON DISEASE IDENTIFICATION AND CLASSIFICATION IN DIFFERENT PLANTS OTHER THAN MAIZE

Apart from prior research, much effort has been expended on artificial intelligence and image processing techniques to identify and categorize a variety of diseases in a variety of plants, including rice, wheat, cotton, tomato, citrus, mango, and banana. This research seeks to create rapid, simple, and effective methods for early detection of plant diseases, as early detection can help lower the risk of plant disease hurting agricultural productivity. Due to the rapid increase in demand for food and agricultural goods as a result of population growth, food security must be assured in order to connect demand and supply; one way to ensuring food security is precision agriculture [36–38]. Precision agriculture, a subset of the third agricultural revolution [39], is a data-driven, technology-enabled approach for sustainable farm management that leverages information communication technologies and Internet of things (IoT) devices to construct agricultural decision support systems [40]. The detection of plant diseases using classic image processing approaches has put forward excellent outcome with a high degree of disease recognition exactness. However, there are some limitations [41], such as the fact that these methods need an indicative portion of time and work, and are extremely subjective to specific conditions and surroundings. Examining the performance of disease rearrangement in complicated contexts is tough; sample sets are tiny and rely heavily on spot segmentation. Numerous studies have been conducted in recent years to identify and characterize plant leaf diseases using image processing techniques combined with artificial intelligence (AI), machine learning (ML), and deep learning (DL) algorithms. Studies [42–45]demonstrate how to detect and classify afflicted plants using traditional image recognition techniques, including K-mean clustering for segmentation, Bayesian discriminant PCA for extracting characteristic parameters, and the usage of a Back Propagation Neural Network (BPNN) model for detection and classification of diseases. The accuracy rates for the models mentioned above are 94.71%, 98.32%, 92.6%, and 97.2%, respectively.

Deep learning technologies are being used to detect diseases in plants more quickly and accurately due to their transparency, the accessibility of larger datasets, efficient graphics processing units (GPUs), training sets, and software libraries such as CUDA, which stand for computing unified device architecture. Numerous approaches leveraging deep learning techniques to recognize and categorize plant diseases have been proposed [46–53] for the reasons outlined previously. The first study [47] uses a three to six-layer CNN classifier to classify three numbers of leguminous plant categories, namely, white bean, soybean, and red bean. In a study by Ma and Kawasaki [49, 50], illnesses in cucumber leaves are identified with an accuracy of 93.4% and 94.9%, respectively, using deep CNN. For the recognition as well as classification of afflicted plants, many methods are used, including the amalgamation of deep learning and transfer learning [51], the use of classifiers such as SVM, K-means, and KNN [52], and a modified deep learning architecture [53].

Apart from the methodologies discussed previously, Goodfellow et al. [54] introduced the global adversarial networks (GANs) model in 2014, which comprises of two components: the generator and the discriminator. The objective of this model is for generation of synthetic samples along with the same attributes as the training distribution. Numerous GAN variants exist, including AR-GAN, CycleGAN, C-DCGAN, DC-GAN, and Leaf GAN. Studies [55–59] have been proposed for the identification of a variety of fruits and their illnesses.

Numerous researchers have proposed novel approaches for detecting plant illnesses using deep learning classifiers in recent years for two reasons: one, the high accuracy, and two, the use of visualization techniques such as visual heat maps and quiet maps to aid in disease detection comprehension. Mohanty et al. [48] classified 14 crop species and 26 plant diseases using both AlexNet and GoogLeNet architectures. The experimental result demonstrates that GoogLeNet performs significantly better than AlexNet. In comparison, Long et al. [60] enhanced AlexNet's performance by combining two types of training, namely, scratch and transfer learning, with an accuracy of approximately 96.53%. Several approaches based on the VGG model and variants of the VGG model have been presented for classifying illnesses in various plants [61–67]. Apart from this research, other efficient approaches for the identification and classification of diseased plants have been developed utilizing deep learning classifiers [68–84] based on LeNet, ResNet, R-CNN, MobileNet, DenseNet, ARNet, and PARNet, respectively. The authors provide an inclusive review of recent researches on recognition and classification of diseased plants using machine learning, deep learning, and IPTs [3, 15, 16, 41, 85–90]. However, this investigation overlooked recent advancements in a number of major DL algorithms used to diagnose and categorize maize leaf diseases.

9.6 CONCLUSION

According to the preceding literature, earlier techniques for identification of diseases in maize were based on the disease's color and shape features. They extract illness signs visible to the naked eye and then use classification methods to identify disorders. These approaches rely largely on competent specialists with a specialized understanding of disease kinds to refine them. Furthermore, the images used in experiments for most recognition tasks are usually obtained in highly demanding circumstances with no disturbance from the outside world. As a result, taking pictures in various environments will negatively impact the test findings. The efficiency of categorization is challenging to assure when novel viruses are identified, irrespective of traditional feature extraction division. Furthermore, the majority of these studies were conducted only to identify and classify maize leaf diseases. Therefore, itis time for researchers to detect, identify, and classify different kinds of diseases in maize plant such as stalk, tassel, and ear to lessen the danger of production loss for corn producers.

REFERENCES

1. Tractor Junction: https://www.tractorjunction.com/blog/top-10-agricultural-producing-countries-in-the-world/, last access 2022/04/13.
2. The Global Economy: https://www.theglobaleconomy.com/rankings/share_of_agriculture/, last access 2022/04/13.
3. Sharma, A., Jain, A., Gupta, P., and Chowdary, V.: Machine learning applications for precision agriculture: A comprehensive review. *IEEE Access* 9, 4843–4873 (2020).

4. Farmer's Portal: https://farmer.gov.in/m_cropstaticsmaize.aspx, last access 2022/04/15.

5. ICAR-Indian Institute of Maize Research: https://iimr.icar.gov.in/india-maze-scenario/, last access 2022/04/15.

6. The CIMMYT Maize Program: *Maize Diseases: A Guide for Field Identification.* 4th edition.CIMMYT (2004).

7. Adotey, N., McClure, A., Raper, T., Florence, R.: Visual symptoms: A handy tool in identifying nutrient deficiency in row crops. The University of Tennessee Institute of Agriculture (2020). Available at: https://news.utcrops.com/2020/07/visual-symptoms-a-handy-tool-in-identifying-nutrient-deficiency-in-row-crops/, last access 2022/04/15.

8. Maize (Corn) – Plant Village: https://plantvillage.psu.edu, last access 2022/04/15.

9. PlantDoc: A Dataset for Visual Plant Disease Detection: https://github.com/pratikkayal/PlantDoc-Dataset, last access 2022/04/15.

10. Corn or Maize Leaf Disease Dataset | Kaggle: https://www.kaggle.com, last access 2022/04/15.

11. Wiesner-Hanks, T., and Brahimi, M.: Image set for deep learning: Field images of maize annotated with disease symptoms. (2019). Retrieved from osf.io/p67rz, last access 2022/04/20.

12. Chaki, J., and Dey, N.: *A Beginner's Guide to Image Preprocessing Techniques.* 1st edition. CRC Press (2018). https://doi.org/10.1201/9780429441134.

13. Sonka, M., Hlavac, V., and Boyle, R.: *Image Processing, Analysis, and Machine Vision.* 4th edition. Cengage Learning (2014). Retrieved from www.cengagebrain.com, last access 2022/04/20.

14. Shorten, C., and Khoshgoftaar, T. M.: A survey on image data augmentation for deep learning. *Journal of Big Data* 6(1), 1–48 (2019).

15. Iqbal, Z., Khan, M. A., Sharif, M., Shah, J. H., Rehman, M. H., and Javed, K.: An automated detection and classification of citrus plant diseases using image processing techniques: A review. *Computers and Electronics in Agriculture* 153, 12–32 (2018).

16. Ferentinos, K. P.: Deep learning models for plant disease detection and diagnosis. *Computers and Electronics in Agriculture* 145, 311–318 (2018).

17. Ishengoma, F. S., Rai, I. A., and Ngoga, S. R.: Hybrid convolution neural network model for a quicker detection of infested maize plants with fall armyworms using UAV-based images. *Ecological Informatics* 67, 101502 (2022).

18. Zeng, W., Li, H., Hu, G., and Liang, D.: Identification of maize leaf diseases by using the SKPSNet-50 convolutional neural network model. *Sustainable Computing: Informatics and Systems* 35, 100695 (2022).

19. Ishengoma, F. S., Rai, I. A., and Ngoga, S. R.: Identification of maize leaves infected by fall armyworms using UAV-based imagery and convolutional neural networks. *Computers and Electronics in Agriculture* 184, 106124 (2021).

20. Xu, Y., Zhao, B., Zhai, Y., Chen, Q., and Zhou, Y.: Maize diseases identification method based on multi-scale convolutional global pooling neural network. *IEEE Access* 9, 27959–27970 (2021).

21. Waheed, A., Goyal, M., Gupta, D., Khanna, A., Hassanien, A. E., and Pandey, H. M.: An optimized dense convolutional neural network model for disease recognition and classification in corn leaf. *Computers and Electronics in Agriculture* 175, 105456 (2020).

22. Panigrahi, K.P., Das, H., Sahoo, A.K., and Moharana, S.C.: Maize leaf disease detection and classification using machine learning algorithms. *Progress in Computing, Analytics and Networking* 1119, 659–669 (2020).

23. Mishra, S., Sachan, R., and Rajpal, D.: Deep convolutional neural network-based detection system for real-time corn plant disease recognition. *Procedia Computer Science* 167, 2003–2010 (2020).

24. Hu, R., Zhang, S., Wang, P., Xu, G., Wang, D., and Qian, Y.: The identification of corn leaf diseases based on transfer learning and data augmentation. *Proceedings of the 2020 3rd International Conference on Computer Science and Software Engineering*, pp. 58–65, IEEE proceedings (2020).

25. Priyadharshini, R. A., Arivazhagan, S., Arun, M., and Mrinalini, A.: Maize leaf disease classification using deep convolutional neural networks. *Neural Computing and Applications* 31, 8887–8895 (2019).

26. Deshapande, A.S., Giraddi, S.G., Karibasappa, K.G., and Desai, S.D.: Fungal disease detection in maize leaves using haar wavelet features. *Information and Communication Technology for Intelligent Systems* 106, 275–286 (2019).

27. Sibiya, M., and Sumbwanyambe, M.: A computational procedure for the recognition and classification of maize leaf diseases out of healthy leaves using convolutional neural networks. *AgriEngineering* 1(1), 119–131 (2019).

28. Aravind, K. R., Raja, P., Mukesh, K. V., Anirudh, R., Ashwin, R., and Szczepanski, C.: Disease classification in maize crop using bag of features and multi-class support vector machine. *2018 2nd International Conference on Inventive Systems and Control (ICISC)*, pp. 1191–1196, IEEE (2018).

29. Lin, Z., Mu, S., Shi, A., Pang, C., and Sun, X.: A novel method of maize leaf disease image identification based on a multi-channel convolutional neural network. *Transaction of the ASABE* 61(5), 1461–1474 (2018).

30. Zhang, X., Qiao, Y., Meng, F., Fan, C., and Zhang, M.: Identification of maize leaf diseases using improved deep convolutional neural networks. *IEEE Access* 6, 30370–30377 (2018).

31. Lv, M., Zhou, G., He, M., Chen, A., Zhang, W., and Hu, Y.: Maize leaf disease identification based on feature enhancement and DMS-robust alexnet. *IEEE Access* 8, 57952–57966 (2020).

32. DeChant, C., Wiesner-Hanks, T., Chen, S., Stewart, E.L., Yosinski, J., Gore, M.A., Nelson, R.J., and Lipson, H.: Automated identification of northern leaf blight-infected maize plants from field imagery using deep learning. *Phytopathology* 107(11), 1426–1432 (2017).

33. Ouppaphan, P.: Corn disease identification from leaf images using convolutional neural networks. *2017 21st International Computer Science and Engineering Conference (ICSEC)*, pp. 1–5, IEEE (2017).

34. Zhang, S.W., Shang, Y.J., and Wang, L.: Plant disease recognition based on plant leaf image. *Journal of Animal and Plant Science* 25(3),42–45 (2015). Retrieved from http://www.thejaps.org.pk/docs/Supplementary/v-25/07.pdf

35. Zhang, L., and Yang, B.: Research on recognition of maize disease based on mobile internet and support vector machine technique. *Advanced Materials Research* 905, 659–662 (2014).

36. Zhang, N., Wang, M., and Wang, N.: Precision agriculture—A worldwide overview. *Computer and Electronics in Agriculture* 36(2–3), 113–132 (2002).

37. Hakkim, V. A., Joseph, E. A., Gokul, A. A., and Mufeedha, K.: Precision farming: The future of Indian agriculture. *Journal of Applied Biology and Biotechnology* 4(6), 68–72 (2016).

38. Gebbers, R., and Adamchuk, V. I.: Precision agriculture and food security. *Science* 327(5967), 828–831 (2010).
39. Stafford, J. V.: Implementing precision agriculture in the 21st century. *Journal of Agriculture Engineering Research* 76(3), 267–275 (2000).
40. Pierce, F. J., and Nowak, P.: Aspects of precision agriculture. In *Advances in Agronomy*, vol. 67, pp. 1–85.Elsevier (1999).
41. Li, L., Zhang, S., and Wang, B.: Plant disease detection and classification by deep learning—A review. *IEEE Access* 9, 56683–56698 (2021).
42. Dubey, S. R., and Jalal, A. S.: Adapted approach for fruit disease identification using images. *Image Processing: Concepts, Methodologies, Tools, and Applications* 1, 1395–1409 (2013).
43. Chai, A. L., Li, B. J., Shi, Y. X., Cen, Z. X., Huang, H. Y., and Liu, J.: Recognition of tomato foliage disease based on computer vision technology. *Acta Horticulturae Sinica* 37(9), 1423–1430 (2010).
44. Li, Z. R., and He, D. J.: Research on identify technologies of apple disease based on mobile photograph image analysis. *Computer Engineering and Design* 31(3095), 3051–3053 (2010).
45. Guan, Z. X., Tang, J., Yang, B. J., Zhou, Y. F., Fan, D. Y., and Yao, Q.: Study on recognition method of rice disease based on image. *Chinese Journal of Rice Science* 24(5), 497–502 (2010).
46. Barbedo, J. G. A.: Factors influencing the use of deep learning for plant disease recognition. *Biosystems Engineering* 172, 84–91 (2018).
47. Grinblat, G. L., Uzal, L. C., Larese, M. G., and Granitto, P. M.: Deep learning for plant identification using vein morphological patterns. *Computers and Electronics in Agriculture* 127, 418–424 (2016).
48. Mohanty, S. P., Hughes, D. P., and Salathe, M.: Using deep learning for image-based plant disease detection. *Frontiers in Plant Science* 7, 1419 (2016).
49. Ma, J., Du, K., Zheng, F., Zhang, L., Gong, Z., and Sun, Z.: A recognition method for cucumber diseases using leaf symptom images based on deep convolutional neural network. *Computers and Electronics in Agriculture* 154, 18–24 (2018).
50. Kawasaki, Y., Uga, H., Kagiwada, S., and Iyatomi, H.: Basic study of automated diagnosis of viral plant diseases using convolutional neural networks. *International Symposium on Visual Computing*, pp. 638–645, Springer (2015).
51. Kessentini, Y., Besbes, M. D., Ammar, S., and Chabbouh, A.: A two-stage deep neural network for multi-norm license plate detection and recognition. *Expert System with Applications* 136, 159–170 (2019).
52. Singh, A. K., Ganapathysubramanian, B., Sarkar, S., and Singh, A.: Deep learning for plant stress phenotyping: Trends and future perspectives. *Trends in Plant Science* 23(10), 883–898 (2018).
53. Saleem, M. H., Potgieter, J., and Arif, K. M.: Plant disease detection and classification by deep learning. *Plants* 8(11), 468–489 (2019).
54. Goodfellow, I.: Nips 2016 tutorial: Generative adversarial networks. arXiv preprint arXiv:1701.00160 (2016).
55. Nazki, H., Yoon, S., Fuentes, A., and Park, D. S.: Unsupervised image translation using adversarial networks for improved plant disease recognition. *Computers and Electronics in Agriculture* 168, 105117 (2020).
56. Tian, Y., Yang, G., Wang, Z., Li, E., and Liang, Z.: Detection of apple lesions in orchards based on deep learning methods of CycleGAN and YOLOV3-dense. *Journal of Sensors* 2019, 7630926 (2019).

57. Hu, G., Wu, H., Zhang, Y., and Wan, M.: A low shot learning method for tea leaf's disease identification. *Computers and Electronics in Agriculture* 163, 104852 (2019).
58. Wu, Q., Chen, Y., and Meng, J.: DCGAN-based data augmentation for tomato leaf disease identification. *IEEE Access* 8, 98716–98728 (2020).
59. Liu, B., Tan, C., Li, S., He, J., and Wang, H.: A data augmentation method based on generative adversarial networks for grape leaf disease identification. *IEEE Access* 8, 102188–102198 (2020).
60. Long, M. S., OuYang, C. J., Liu, H., and Fu, Q.: Image recognition of camellia oleifera diseases based on convolutional neural network & transfer learning. *Transaction of the Chinese Society of Agricultural Engineering* 34(18), 194–201 (2018).
61. Lu, J., Hu, J., Zhao, G., Mei, F., and Zhang, C.: An in-field automatic wheat disease diagnosis system. *Computers and Electronics in Agriculture* 142, 369–379 (2017).
62. Ha, J. G., Moon, H., Kwak, J. T., Hassan, S. I., Dang, M., Lee, O. N., and Park, H. Y.: Deep convolutional neural network for classifying Fusarium wilt of radish from unmanned aerial vehicles. *Journal of Applied Remote Sensing* 11(4), 042621 (2017).
63. Xu, J. H., Shao, M. Y., Wang, Y. C., and Han, W. T.: Recognition of corn leaf spot and rust based on transfer learning with convolutional neural network. *Transactions of the Chinese Society for Agricultural Machinery* 51(2), 230–236 (2020).
64. Li, K. Z., Lin, J. H., Liu, J. R., and Zhao, Y. D.: Using deep learning for image-based different degrees of ginkgo leaf disease classification. *Information* 11(2), 95 (2020).
65. Ren, S. G., Jia, F. W., Gu, X. J., Yuan, P. S., Xue, W., and Xu, H. L.: Recognition and segmentation model of tomato leaf diseases based on deconvolution-guiding. *Transaction of the Chinese Society of Agricultural Engineering* 36(2), 186–195 (2020).
66. Chen, J., Chen, J., Zhang, D., Sun, Y., and Nanehkaran, Y. A.: Using deep transfer learning for image-based plant disease identification. *Computers and Electronics in Agriculture* 173, 105393 (2020).
67. Zhen, W., Shanwen, Z., and Baoping, Z.: Crop diseases leaf segmentation method based on cascaded convolutional neural network. *Computer Engineering and Applications* 56(15), 242–250 (2020).
68. Cruz, A. C., Luvisi, A., De-Bellis, L., and Ampatzidis, Y.: Vision-based plant disease detection system using transfer and deep learning. *2017 Asabe Annual International Meeting* Vol. 1, American Society of Agricultural and Biological Engineers (2017).
69. Kerkech, M., Hafiane, A., and Canals, R.: Deep learning approach with colorimetric spaces and vegetation indices for vine diseases detection in UAV images. *Computers and Electronics in Agriculture* 155, 237–243 (2018).
70. Jiang, F., Li, Y., Yu, D. W., Min, S., and Zhang, E.: Design and experiment of tobacco leaf grade recognition system based on caffe. *Journal of Chinese Agricultural Mechanization* 40(1), 126–131 (2019).
71. Picon, A., Alvarez-Gila, A., Seitz, M., Ortiz-Barredo, A., Echazarra, J., and Johannes, A.: Deep convolutional neural networks for mobile capture device-based crop disease classification in the wild. *Computers and Electronics in Agriculture* 161, 280–290 (2019).

72. Qiu, R., Yang, C., Moghimi, A., Zhang, M., Steffenson, B. J., and Hirsch, C. D.: Detection of fusarium head blight in wheat using a deep neural network and color imaging. *Remote Sensing* 11(22), 2658 (2019).

73. Thapa, R., Snavely, N., Belongie, S., and Khan, A.: The plant pathology2020 challenge dataset to classify foliar disease of apples. *Applications in Plant Sciences* 8(9), e11390 (2020).

74. He, X., Li, S. Q., and Liu, B.: Grape leaf disease identification based on multi-scale residual network. *Computer Engineering* 47(5), 285–291, 300 (2021).

75. Yu, X. D., Yang, M. J., Zhang, H. Q., Li, D., Tang, Y. Q., and Yu, X.: Research and application of crop diseases detection method based on transfer learning. *Transaction of the Chinese Society of Agricultural Engineering* 51(10), 252–258 (2020).

76. Esgario, J. G. M., Krohling, R. A., and Ventura, J. A.: Deep learning for classification and severity estimation of coffee leaf biotic stress. *Computers and Electronics in Agriculture* 169, 105162 (2020).

77. Arora, K., Kumar, A., Kamboj, V.K., Prashar, D., Jha, S., Shrestha, B., and Joshi, G.P.: Optimization methodologies and testing on standard benchmark functions of load frequency control for interconnected multi area power system in smart grids. *Mathematics* 8, 980 (2020).

78. Li, X. R., Li, S. Q., and Liu, B.: Apple leaf disease detection method based on improved faster R-CNN. *Computer Engineering* 47(11), 298–304 (2021).

79. Ozguven, M. M., and Adem, K.: Automatic detection and classification of leaf spot disease in sugar beet using deep learning algorithms. *Physica A: Statistical Mechanics and Its Applications* 535, 122537 (2019).

80. Zhou, M. M.: Apple foliage diseases recognition in Android system with transfer learning-based. M.S. thesis, Dept. Inf. Eng., Northwest A&F Univ., Yangling, China (2019).

81. Liu, Y., Feng, Q., and Wang, S. Z.: Plant disease identification method based on light-weight CNN and mobile application. *Transaction of the Chinese Society of Agricultural Engineering* 35(17), 194–204 (2019).

82. Arora, K., Kumar, A., Kamboj, V. K., Prashar, D., Shrestha, B., and Joshi, G. P.: Impact of renewable energy sources into multi area multi-source load frequency control of interrelated power system. *Mathematics* 9, 186 (2021).

83. Nagasubramanian, K., Singh, A. K., Singh, A., Sarkar, S., and Subramanian, B. G.: Usefulness of interpretability methods to explain deep learning-based plant stress phenotyping. *Computer Science*, arXiv preprint arXiv: 2007.05729 (2020).

84. Hu, Z. W., Yang, H., Huang, J. M., and Xie, Q. Q.: Fine-grained tomato disease recognition based on attention residual mechanism. *Journal of South China Agricultural University* 40(6), 124–132 (2019).

85. Kamilaris, A., and Prenafeta-Boldu, F. X.: Deep learning in agriculture: A survey. *Computers and Electronics in Agriculture* 147, 70–90 (2018).

86. Ngugi, L. C., Abelwahab, M., and Abo-Zahhad, M.: Recent advances in image processing techniques for automated leaf pest and disease recognition— A review. *Information Processing in Agriculture* 8(1), 27–51 (2020).

87. Singh, V., Sharma, N., and Singh, S.: A review of imaging techniques for plant disease detection. *Artificial Intelligence in Agriculture* 4, 229–242 (2020).

88. Jadhav, S., and Garg, B.: Comprehensive review on machine learning for plant disease identification and classification with image processing. *Proceedings of International Conference on Intelligent Cyber-Physical Systems*, pp. 247–262, IEEE proceedings (2022).

89. Sethy, P. K., Pandey, C., Sahu, Y. K., and Behera, S. K.: Hyperspectral imagery applications for precision agriculture-a systemic survey. *Multimedia Tools and Applications* 1, 1–34 (2021).

90. Sethy, P. K., Barpanda, N. K., Rath, A. K., and Behera, S. K.: Image processing techniques for diagnosing rice plant disease: A survey. *Procedia Computer Science* 167, 516–530 (2020).

Chapter 10

Low-power architectural design and implementation of reconfigurable data converters for biomedical application

P. Rama Krishna
Anurag University, Hyderabad, India

T. Santosh Kumar
CMR Institute of Technology, Hyderabad, India

Sandhya Avasthi
ABES Engineering College, Gaziabad, India

Suman Lata Tripathi
Lovely Professional university, Phagwara, India

B. Srikanth
Vardhaman College of Engineering, Hyderabad, India

Kakarla Hari Kishore
Koneru Lakshmaiah Education Foundation, Guntur, India

CONTENTS

DOI: 10.1201/9781003407409-10

10.1 INTRODUCTION

A subsystem that would be capable of managing a variety of ADCs as well as DACs with varying speeds and desired outcomes would be necessary for plan for test (DFT) for CMOS mixed signal circuit test methods. Additionally, a subsystem, which is capable of identifying and manipulating a signal (bio-impedance) which is close because of appropriateness and then also rehash assortment, is necessary for bio-impedance checking system. A few mixed signal systems remain open, but they all have either an ADC or perhaps a DAC, but not both. Throughout this chapter, a design for a low-power reconfigurable data converter is presented. With the aid of an imaginative internal trading structure, it is possible to arrange it as an ADC & DAC. Following that, an ambitious tailored adaptation circuitry with low power consumption is intended to design this conversion.

Traditionally, pipelined ADCs have been deemed the optimal design for high-bandwidth (BW) applications requiring a resolution of 10–15 bits. Nevertheless, typical SC flip-around MDACs continue to pose design issues, like inter-stage gain fluctuations as well as capacitive mismatches between the MDACs, which restrict the total conversion performance. Similarly, SC-MDACs often need a high-power consumption from the residue amplifier because of their high capacitive loading, short settling time and low feedback factor, which would be the case whenever multiple bits are being resolved at each stage [1–6]. Additionally, extensive research has been concentrated on even the most energy-intensive component of pipeline ADCs, namely, residue amplifier. Lim and Flynn [7] use a ring amplifier that operates mostly on dynamic power rather than the high static current required by typical OPAMPs. Pipelined ADCs based on current-mode (CM) MDACs having previously been presented as a means of improvement in the power consumption of SC MDACs. For example, Chevella [8] pioneered current-mode design in BiCMOS technology, wherein current would be injected just at inverted amplifier's imaginary ground node. Due to high gain installed in the inverting amplifier, the feedback factor has been modest; hence, the residual amplifier continues to play a significant role in the MDAC's power consumption. Additionally, pipelined designs employing switching signal for residues synthesis have been reported [9]. These architectures include CM MDACs as well as continuous-time residue amplifiers [14].

This adaptable circuit may be compelled to adjust its scanning rate and goals in response to the data signal's quantity and repetition. The data converter may be configured as simple periodic voltage generators and uti-lised as test-redesign generator for CMOS BIST. By inoculating a sine sam-ple, it is possible to avoid tissue damage caused by these artefacts. The impedance of a bodily tissue containing significant biological data describes its architecture; plasma has a frequency band of 1 Hz to many MHz as well as an amplitude ranging from 15 uA to 2 mA. A diverse array of progressing components in the impedance of bio-objects would be favoured in the pro-cedure, as is a lower current distribution and changed tactile conceptual model activity. To collect, process and send this critical information, implant-able devices are used, which are confined by limited space and power, among other factors. Due to the fact that reconfigurable data converters have enough potential to develop control capabilities for structures with chang-ing speed and desired goals, there is a greater emphasis on these reconfigu-rable converters. These are very endearing, particularly for implantable bio-impedance sensing devices applications within CMOS systems.

10.2 SYSTEM ARCHITECTURE

A square graph of the proposed framework is shown in Figure 10.1. It com-prises of a variety of indistinguishable information transformation stages, and an interconnection system (ICS) every datum change organize (DCS) comprises of an intensifier, pipelined ADC in ADC block and DACs in DAC block. The proposed data converter has been designed on Cadence Virtuso version 6.5 at 180 nm technology node.

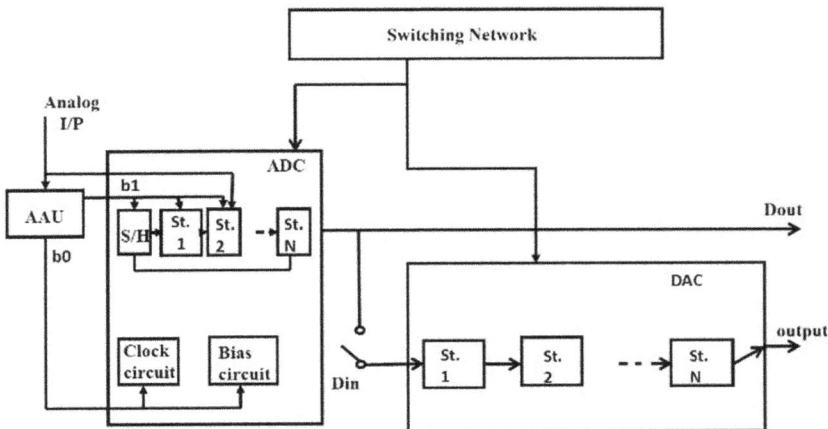

Figure 10.1 Reconfigurable data converter.

10.3 ANALOG-TO-DIGITAL CONVERTER

The architecture is meant to use a highly capable ADC. It should not only be able to modify its inspection speed as well as resolution, but it should also be able to promptly and consistently determine whichever design should work in to comply with the basic information signal. We developed a low-power, low-complexity programmed adjusting device. It is a simple circuit that is appropriate for assessing biomedical as well as mixed signal circuits. It customizes this device with an appropriate method of operation based on signal frequency and magnitude. Figure 10.2 presents a pipelined ADC.

An ADC converts a continuously amplitude analogue signal into a digital signal. The conversion process requires quantization of data, which inherently introduces a tiny level of noise/error. In contrast, an ADC conducts conversions on a periodic basis, sampling incoming data, thus lowering the allowable spectrum of the input signal, rather than continuously. The ADC's bandwidth and S/N ratios are essentially defined by their efficiency. The ADC's bandwidth would be dictated by its samples per second. The SNR of ADC does have a number of implications. The SNR of an ADC is very often important in terms of its own effective number of bits (ENOB), and the best ADC does have an ENOB equal to a solution. Numerous ADC designs have been developed to satisfy a variety of needs. Several of them are significant in industry and other sectors: SAR-ADC, pipeline ADC and Flash ADC.

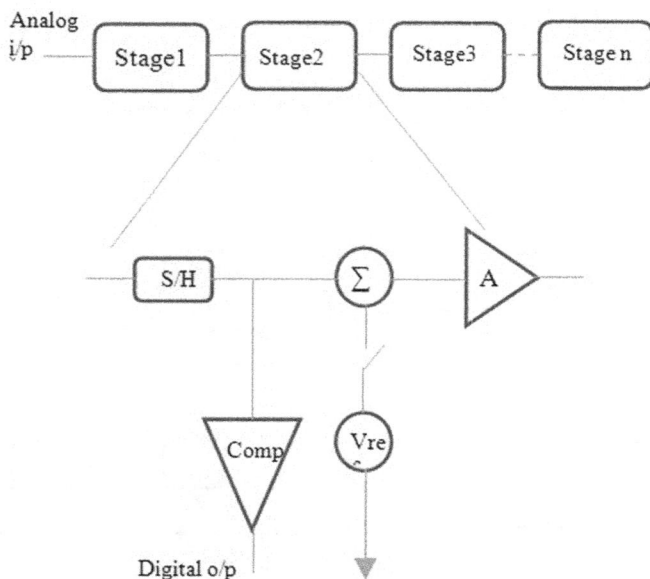

Figure 10.2 Pipeline ADC.

Usually, the SAR ADC changes significantly with the sampling frequency [15–18]. This is also why pipeline ADC would be expected to minimize energy consumption via self-adaptation. This technology transmits the incoming signal directly toward the converter and automated correction unit. Fundamentally, there really are two ways to configure the resolution: by disabling the start or end phases. A N-step ADC pipeline presented here is a 1bit converter per stream. The ADC pipeline has been composed of N stages connected in series (Figure 10.2), each of which is capable of producing high-resolution (8 to 12-bit) data at a reasonable rate. Every stage has a sample and hold circuit, a subtracting circuit, a residual amplifier, as well as a gate switch. Every step of the conversion follows the same procedure: Once input data has been filtered, connect this signal to $V_{ref/2}$. The result of every comparator would be the bit change for such step. Whenever the output of V_m exceeds $V_{ref/2}$ (output is 1), then held signal's $V_{ref/2}$ is being removed and passed to the amplifier. Within case of Vin $V_{ref/2}$ (output is 0), the amplifier receives the original input signal. The residue is generally defined as the output of the converter's various stages. Multiply this result by two and move it to the subsequent stage sample and hold circuit. The primary benefit of the pipeline converter has been its excellent performance. After the initial delay of N clock cycles, every clock cycle conversion occurs. Even though the second stage handles the leftover of the first stage, the very first stage remains accessible for future samples. Every step recycles the remnants of the preceding stage, enabling for faster conversions.

10.4 AUTO ADAPTION UNIT

Digital control resolution and sample rate configurations, or both, are possible. Numerous systems depend on manual signals from the user or on various customizable controllers to effect these changes. This is a much more powerful engine with increased strength. Rather than that, the ADC configuration would be decided instantly, without the need to consult a third-party source. An auto adaptation unit's circuit diagram is seen in Figure 10.3. It is a small, low-power design that is well-suited for bio-impedance monitoring devices and CMOS circuit testing. The biological impedance sensors send signals directly to the conversion and auto adaptation unit. Control bits, q0 & q1, are utilised to set the device there in appropriate operating mode based on the signal frequency and amplitude.

10.4.1 Sampling rate control signal

To modify the sampling frequency, a Schmitt trigger would be utilised to convert the enhanced input signal to a square wave. After that, an FVC will be utilised to generate a DC voltage that is inversely proportional to the frequency of the input signal. The result of the FVC usually compared with

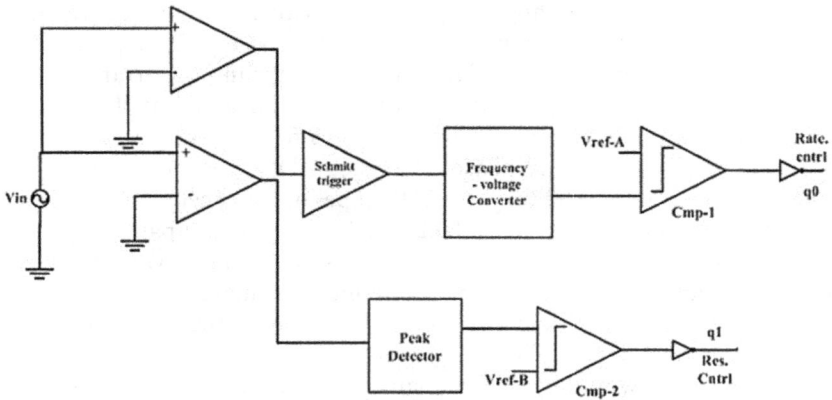

Figure 10.3 Auto adaption unit.

the corresponding voltage. The output comprises buffered and functions as a control bit for converter's sampling frequency.

10.4.2 Resolution control signal

The amplitude of the incoming signal is obtained using an envelope detector using CMOS technology. The envelope of the input has been monitored and compared with either a reference voltage using only a simple CMOS envelop detector. Every block act as a current bias within automated adapter unit to ensure steady functioning.

10.5 DIGITAL-TO-ANALOGUE CONVERTER

DAC is indeed an acronym for digital data collection and processing system using analogue signals. A CPU, ASIC, /FPGA could be used to create the digital data. There are numerous DAC designs, and the usefulness of a DAC for a particular application would be determined by its figure of merit, which includes resolution and maximal sampling frequency. Because the DAC has the ability to degrade a signal, it is recommended to specify a DAC with minor implementation defects. The audio DAC operates at a high resolution and low frequency, whereas the video DAC operates at a low to moderate resolution and high frequency [19–22]. A current steering DAC is composed of two blocks: a current mirror circuitry with switches, in which all the MSB & LSB widths remain identical for constant current transistors. Figure 10.4 depicts the suggested CS-DAC design in its entirety. Main building blocks in this DAC are block switches, current array sources, biasing voltages and filter. The current from across all eight array inputs has been routed through these blocks switches as iout+ & iout– outputs. Additionally, the binary input signal would be monitored. An On functionality of the DAC's branching

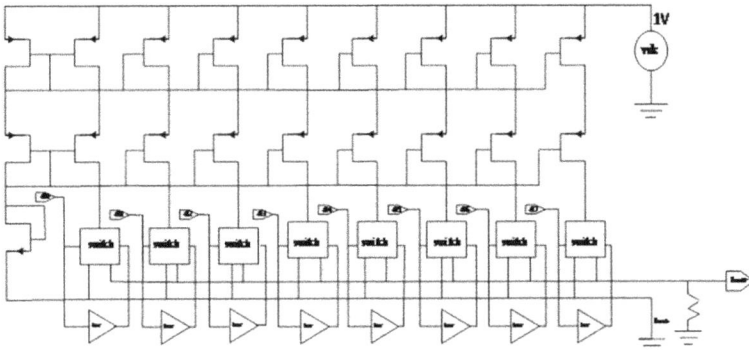

Figure 10.4 Schematic of CS-DAC.

may be handled using these block switches. The changing times are close to that of the current moving through the branch, which increases the complex characteristics. In this situation, the conversion time would be inversely proportional to the MOS transistor dimensions. The LSB corresponding MOS switch is quite small, but the MSB equivalent is 128 times bigger.

10.6 PERFORMANCE PARAMETERS

For the efficiency of the system, DACs are very necessary. Such systems are primarily distinguished by:

10.6.1 Static performance

Differential nonlinearity (DNL) quantifies how far apart adjacent values using two analogue codes diverge from the optimal 1 LSB stage. Integral nonlinearity (INL) demonstrates how the DAC transfer characteristic deviates from the ideal. Usually, the ideal functionality is just a straight line; INL illustrates how the actual voltage in LSBs (1 LSB step) over a certain code value differs from it. Passive elements such as resistance are primarily limited to the thermal noise produced by gain, offset and noise. This noise generalised for recording applications and at ambient temperatures is slightly less than 1 μV (microvolt). The 'spurious free dynamic range' is one in which spur is a part of harmonic distortion. The Signal to Noise and Distortion ration expressed in dB, the ratio of the power of the transformed main signal to the amount of noise and the harmonic spurs are produced. The ith harmonic distortion (HDi) represents the power of ith harmonic of the transformed main signal. The total harmonic distortion (THD) is the cumulative intensity of all HDi. The D/A converter assumes that monotonic of the peak DNL error is greater than 1LSB, although a maximum DNL of 1 LSB can be found in many monotonic converters.

10.7 INTERCONNECTION SYSTEM (ICS)

The switch network (ICS) assists in organizing the data converter system into the appropriate mode of operation, which would be to configure like an ADC/DAC, as well as in selecting the DAC's resolution. To determine the ADC's resolution, an auto adaptation circuit is implemented; the auto adaption circuit was mentioned before in this section.

10.8 DTMOS LOGIC

In the last few years, even as market for handy devices like as phones, portable PCs and some other low-power implementations has expanded, the design of simple circuits that need minimal power and low voltage has been a serious concern. The limit voltage has been one of the constraints on the use of tiny devices as well as the construction of many other low-power circuitry operating at low voltage. As a result, lowering the limit voltage is critical for low-control, low-voltage activities. The DTMOS approach [23] is the optimal solution for lowering the limit voltage (Vth). In this regard, an effective strategy for reducing energy use is to lower the dc power supply (V_{dd}). Thus, the reduction in the power supply voltage (V_{dd}) would be dependent on one of the factors, which is the limit voltage [24]. Thus, one possible arrangement is to implement CMOS transistors having dynamic Vth, which would be the underlying concept of the DTMOS method. While in the 'off' state, the DTMOS transistor has a high edge signature to restrict spilling current, yet, whenever in the 'on' state, it operates as a low-edge device at reduced supply voltage for large current driving capability. It is one of the factors that contributes to the DTMOS method's suitability for reduced low-power systems. In strong limit CMOS (DTMOS), the edge potential is gradually changed to meet the circuit's operating conditions. This is shown in Figure 10.5, where the NMOS & PMOS transistor bodies are forcefully

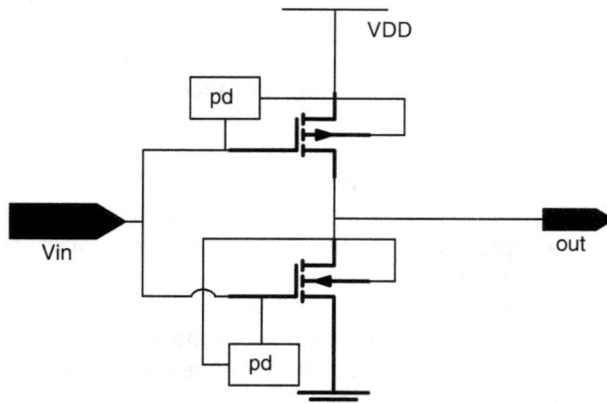

Figure 10.5 DTMOS-based circuit.

one-sided. The potential dividers (pd) connected with the inverter provide an enough body biassing voltage toward both PMOS & NMOS transistors. In standby mode, a high-edge voltage results in reduced leakage current, whereas a low-edge voltage plays an important part in the dynamic mode of operation. The DTMOS voltage would be constrained by the diode used in bulk silicon innovation. A p-n diode between the source as well as the body should be flipped one-sided. As a result, this technology is appropriate for extreme low voltage (0.6 V or less) circuits in bulk CMOS [25–27].

The DTMOS approach decreases both the transistor's off-state leakage current and the threshold voltage, while the transistors stand in the on-state (VBS > 0).

10.9 RESULTS AND DISCUSSION

Figure 10.6 illustrate the DNL and INL.

$$DNL = \left| \left[\left(V_{D+1} - V_D \right) / V_{LSB-IDEAL} - 1 \right] \right|, where\, 0 < D < 2^N - 2 \qquad (10.1)$$

V_D denotes the physical value associated with digital output coding D, N denotes the ADC resolution, whereas VLSB-Optimum denotes the ideal spacing between two consecutive digital codes. Increased values of DNL often restrict the ADC's results in terms of SNR and SFDR by adding noise and erroneous components beyond the impacts of quantization. Here, we achieved low DNL of ±0.6 LSB.

$$INL = \left| \left[\left(V_D - V_{ZERO} \right) / V_{LSB-IDEAL} \right] - D \right| \qquad (10.2)$$

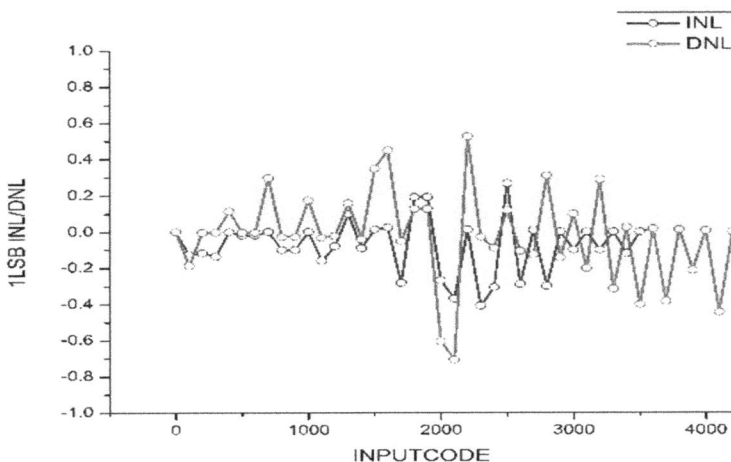

Figure 10.6 INL/DNL plot of DAC.

where

D lies in range of $0 < D < 2^N-1$.

V_D – analogue value representing digital output code D,

N-ADC's resolution,

V_{ZERO} – minimum analogue input corresponding to various all-zero output code,

$V_{LSB-IDEAL}$ – ideal spacing for various two output codes.

INL of ±0.3LSB govern that DAC has robust matching.

SNR as well as SINAD values of 71.06 and 72.3 dB for 12-bit and 8-bit mode, respectively, are shown in Figures 10.7 and 10.8. The blue colour line marked by 'F' denotes basic power, whereas the orange lines represent both SNR & SINAD noise. Figure 10.9 illustrates square boxes containing numbers indicating the harmonics that have been displayed.

The computed SFDR of 81.85 and 82.2 dB for 12-bit and 8-bit mode are shown in Figure 10.9. The area shaded in blue denotes SFDR, where 'F' denotes basic power and 'S' denotes spurs.

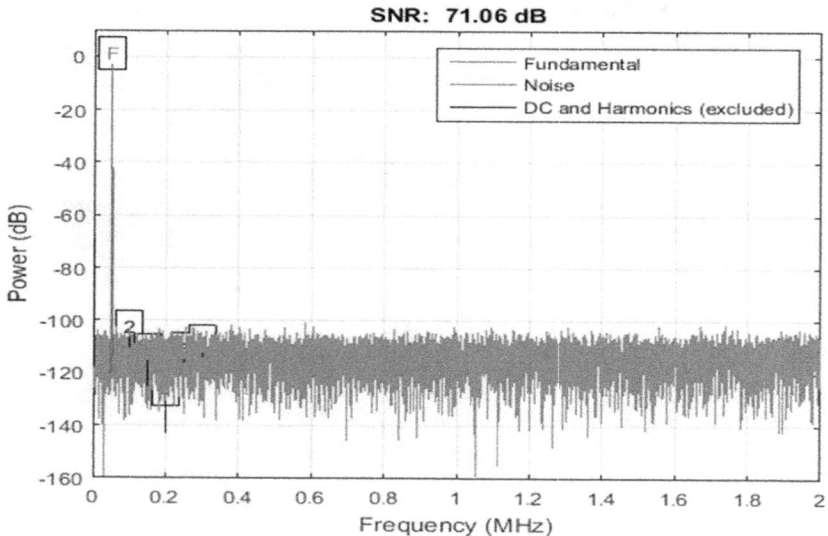

Figure 10.7 SNR plot.

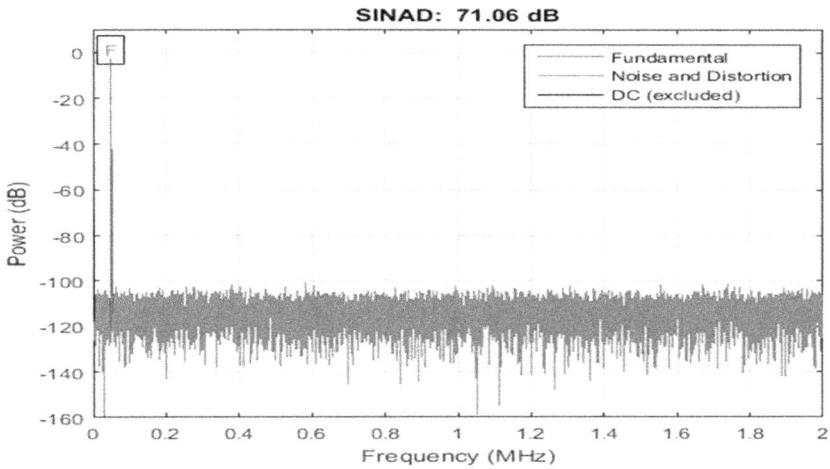

Figure 10.8 SINAD plot.

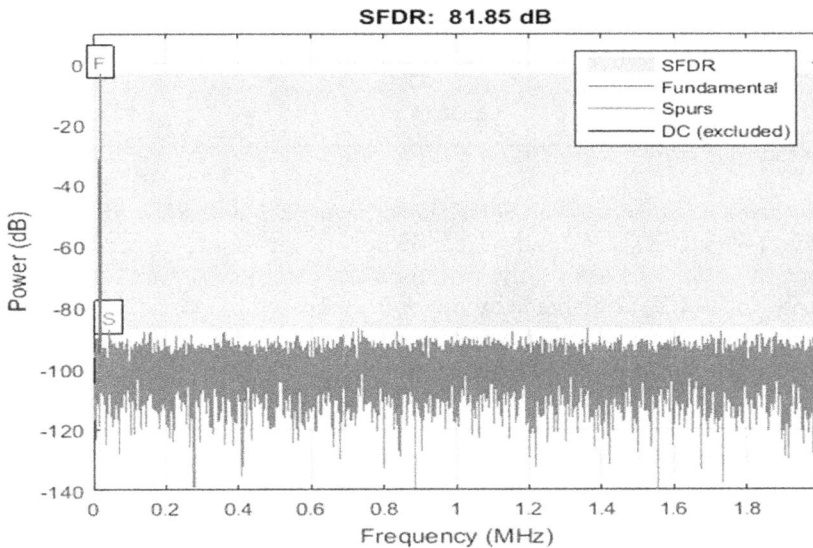

Figure 10.9 SFDR plot.

10.10 PERFORMANCE

Tables 10.1 and 10.2 show the comparison of the proposed ADC and DAC with current data converters in terms of performance. The recommended

Table 10.1 Comparison of ADC in this work with previous studies

Parameter	[6]	[1]	This work
Technology	180 nm	130 n	180 nm
Supply voltage	1.8	1.5	1V
Sampling rate	10–80	0.1–20	10 ks/s – 1 Gs/s
Resolution	10	8/10	8/12
Power	81–94 mW	86 uW/1m	214.70/844.70 uW
SINAD (dB)	58.8–56.5	48.2/58.7	72.3/71.06
SNR (dB)	—	—	72.3/71.06
SFDR (dB)	—	—	82.2/81.85

Table 10.2 Comparison of DAC in this work with previous studies

Parameter	[13]	[11]	This work
Technology (nm)	180	180	180
Supply voltage (V)	1.8	1.8	1
Resolution	14	10	8/12
Power (W)	97 m	4 m	99/557 u
INL	±0.6 LSB	<±0.3 LSB	±0.3 LSB
DNL	±0.56 LSB	±0.5 LSB	±0.6 LSB

converter's SNR as well as SFDR have been 72.3 and 82.2 dB for 8-bit mode, as well as 71.06 and 81.85 dB for 12-bit mode, respectively. The result in Table 10.1 shows the proposed ADC similar resolution at lower supply node (1 V) at higher sampling rate compared to existing work. ADC block prosed can be further extended with new techniques of current mirrors and comparator blocks [28, 29].

10.11 CONCLUSION

This study introduces a low-power reconfigurable ADC as well as DAC. It was accomplished by using DTMOS logic and several transistors in sub-threshold regions, as well as by configuring the ADC to operate in the right mode for objectives and sampling rate for acquired data. The ADC features a synchronized adaptation unit. The additional power consumption as well as area required by the adaptation unit and switching system are negligible; a bio-impedance framework is used to illustrate an impedance above the wide recurrence range with increased productive output, as well as they could also be used for those specific applications, as they can be intended as an ADC or DAC depending on the application and thus withstand very high sampling rates up to 1 Gs/s.

REFERENCES

1. T. C. Randall, I. Mahbub and S. K. Islam, (2015) "Reconfigurable Analog-to-Digital Converter for Implantable Bio-impedance Monitoring," in *IEEE Biowireless*, vol. 1, pp. 32–34.
2. S. Sengupta and M. L. Johnston, (2021) "A Widely Reconfigurable Piecewise-Linear ADC for Information-Aware Quantization," in *IEEE Transactions on Circuits and Systems II: Express Briefs*, vol. 68, no. 4, pp. 1073–1077, doi: 10.1109/TCSII.2020.3031262.
3. M. Elsobky, A. Mohamed, T. Deuble, J. Anders and J. N. Burghartz, (2021) "A 12-to-15 b, 100-to-25 kS/s Resolution Reconfigurable, Power Scalable Incremental ADC Using Ultrathin Chips," in *IEEE Sensors Letters*, vol. 5, no. 2, pp. 1–4, Art no. 5500104, doi: 10.1109/LSENS.2021.3051259.
4. U. Hazarika, M. Das and K. K. Sarma, (2020) "A Reconfigurable Time Multiplexed Multichannel ADC Model for Efficient Data Acquisition," in *2020 International Conference on Computational Performance Evaluation (ComPE)*, pp. 809–812, doi: 10.1109/ComPE49325.2020.9200007.
5. K. Arora, A. Kumar, V. K. Kamboj, D. Prashar, S. Jha, B. Shrestha and G. P. Joshi, (2020) "Optimization Methodologies and Testing on Standard Benchmark Functions of Load Frequency Control for Interconnected Multi Area Power System in Smart Grids," in *Mathematics*, vol. 8, p. 980.
6. J. Adsul, P. P. Vaidya and J. M. Nair, (2016) "A New Method of Reconfigurable ADC Using Calibrated Programmable Slopes," in *2016 International Conference on Communication and Electronics Systems (ICCES)*, pp. 1–6, doi: 10.1109/CESYS.2016.7889985.
7. I. S. Kim, K. Lee and M. Lee, (2018) "Modeling Random Clock Jitter Effect of High-Speed Current-Steering NRZ and RZ DAC," in *IEEE Transactions on Circuits and Systems I: Regular Papers*, vol. 65, no. 9, pp. 2832–2841, doi: 10.1109/TCSI.2018.2821198.
8. S. Chevella, D. O'Hare and I. O'Connell, (2020) "A Low-Power 1-V Supply Dynamic Comparator," in *IEEE Solid-State Circuits Letters*, vol. 3, pp. 154–157, doi: 10.1109/LSSC.2020.3009437.
9. T.-C. Hung, F.-W. Liao and T.-H. Kuo, (2019) "A 12-Bit Time-Interleaved 400-MS/s Pipelined ADC with Split-ADC Digital Background Calibration in 4,000 Conversions/Channel," in *IEEE Transactions on Circuits and Systems II: Express Briefs*, vol. 66, no. 11, pp. 1810–1814, doi: 10.1109/TCSII.2019.2895694.
10. K. Arora, A. Kumar, V. K. Kamboj, D. Prashar, B. Shrestha, G. P. Joshi, 2021 "Impact of Renewable Energy Sources into Multi Area Multi-Source Load Frequency Control of Interrelated Power System," in *Mathematics*, vol. 9, p. 186.
11. J. Mao, M. Guo, S. Sin and R. P. Martins, (2018) "A 14-Bit Split-Pipeline ADC with Self-Adjusted Opamp-Sharing Duty-Cycle and Bias Current," in *IEEE Transactions on Circuits and Systems II: Express Briefs*, vol. 65, no. 10, pp. 1380–1384, doi: 10.1109/TCSII.2018.2851944.
12. C. Briseno-Vidrios et al., (2018) "A 44-fJ/Conversion Step 200-MS/s Pipeline ADC Employing Current-Mode MDACs," in *IEEE Journal of Solid-State Circuits*, vol. 53, no. 11, pp. 3280–3292, doi: 10.1109/JSSC.2018.2863959.
13. V. Sarma, N. A. Jacob, B. D. Sahoo, V. Narayanaswamy and V. Choudhary, (2018) "A 250-MHz Pipelined ADC-Based $f_{S}/4$ Noise-Shaping Bandpass

ADC," in *IEEE Transactions on Circuits and Systems I: Regular Papers*, vol. 65, no. 6, pp. 1785–1794, doi: 10.1109/TCSI.2017.2766883.

14. S. Choi et al., (2017) "A Self-Biased Current-Mode Amplifier with an Application to 10-Bit Pipeline ADC," in *IEEE Transactions on Circuits and Systems I: Regular Papers*, vol. 64, no. 7, pp. 1706–1717, doi: 10.1109/TCSI.2017.2676105.

15. C. Erdmann et al., (2015) "A Heterogeneous 3D-IC Consisting of Two 28 nm FPGA Die and 32 Reconfigurable High-Performance Data Converters," in *IEEE Journal of Solid-State Circuits*, vol. 50, no. 1, pp. 258–269, doi: 10.1109/JSSC.2014.2357432.

16. C.-C. Liu and M.-C. Huang, (2017) "A 0.46 mW 5MHz-BW 79.7 dB-SNDR Noise-Shaping SAR ADC with Dynamic-Amplifier-Based FIR-IIR Filter," in *IEEE International Solid-State Circuits Conference. Digest of Technical Papers*, pp. 466–467, USA.

17. Y. Zhang, C.-H. Chen, T. He, K. Sobue, K. Hamashita and G. C. Temes, (2017) "A Two-Capacitor SAR-Assisted Multi-Step Incremental ADC with a Single Amplifier Achieving 96.6 dB SNDR over 1.2 kHz BW," in *Proc. of IEEE Custom Integrated Circuits Conference (CICC)*, Austin, TX, USA, pp. 1–4.

18. S. Karmakar, B. Gönen, F. Sebastiano, R. V. Veldhoven and K. A. A. Makinwa, (2018) "A 280 μW Dynamic-Zoom ADC with 120 dB DR and 118 dB SNDR in 1 kHz BW," in *IEEE International Solid-State Circuits Conference. Digest of Technical Papers*, February, pp. 238–239, China.

19. S. Mehta, D. O'Hare, V. O'Brien, E. Thompson and B. Mullane, (2018) "Analysis and Design of a Tri-Level Current-Steering DAC with 12-Bit Linearity and Improved Impedance Matching Suitable for CT-ADCs," in *IEEE Open Journal of Circuits and Systems*, vol. 1, pp. 34–47, doi: 10.1109/OJCAS.2020.2994838.

20. G. Park and M. Song, (2015) "A CMOS Current-Steering D/A Converter with Full-Swing Output Voltage and a Quaternary Driver," in *IEEE Transactions on Circuits and Systems II: Express Briefs*, vol. 62, no. 5, pp. 441–445, doi: 10.1109/TCSII.2014.2386259.

21. E. Bechthum, G. I. Radulov, J. Briaire, G. J. G. M. Geelen and A. H. M. van Roermund, (2016) "A Wideband RF Mixing-DAC Achieving IMD < -82 dBc Up to 1.9 GHz," in *IEEE Journal of Solid-State Circuits*, vol. 51, no. 6, pp. 1374–1384, doi: 10.1109/JSSC.2016.2543703.

22. S. Kim, M. Kim, B. Sung, H. Kang, M. Cho and S. Ryu, (2015) "A SUC-Based Full-Binary 6-Bit 3.1-GS/s 17.7-mW Current-Steering DAC in 0.038 mm," in *IEEE Transactions on Very Large Scale Integration (VLSI) Systems*, vol. 24, no. 2, pp. 794–798, doi: 10.1109/TVLSI.2015.2412657.

23. H. Homulle, L. Song, E. Charbon and F. Sebastiano, (2018) "The Cryogenic Temperature Behavior of Bipolar, MOS, and DTMOS Transistors in Standard CMOS," in *IEEE Journal of the Electron Devices Society*, vol. 6, pp. 263–270, doi: 10.1109/JEDS.2018.2798281.

24. Y.-K. Teh and P. K. T. Mok, (2016) "DTMOS-Based Pulse Transformer Boost Converter with Complementary Charge Pump for Multisource Energy Harvesting," in *IEEE Transactions on Circuits and Systems II: Express Briefs*, vol. 63, no. 5, pp. 508–512, doi: 10.1109/TCSII.2015.2505259.

25. K. Kim and S. Kim, (2015) "Design of Schmitt Trigger Logic Gates Using DTMOS for Enhanced Electromagnetic Immunity of Subthreshold Circuits," in *IEEE Transactions on Electromagnetic Compatibility*, vol. 57, no. 5, pp. 963–972, doi: 10.1109/TEMC.2015.2427992.
26. P. Ramakrishna and K. H. Kishore, (2018) "Design of an Ultra-Low Power CMOS Comparator for Data Converters," in *Journal of Advanced Research in Dynamical and Control Systems*, vol. 1, pp. 1347–1352.
27. G. T. Varghesea and K. Mahapatra, (2016) "A Low Power Reconfigurable Encoder for Flash ADCs," in *Elsevier, Open Access Article, Procedia Technology*, vol. 25, pp. 574–581.
28. T. Singh and S. L. Tripathi, 2021 "Design of a 16 Bit 500 MS/s SAR ADC for Low Power Application," in *Electronic Device and Circuits Design Challenges to Implement Biomedical Applications*. Elsevier, doi: 10.1016/B978-0-323-85172-5.00021-6.
29. T. Singh and S. L. Tripathi, 2022 "An Efficient Approach to Design a Comparator for SAR-ADC," in *2022 IEEE VLSI Device Circuit and System (VLSI DCS)*, pp. 116–122, doi: 10.1109/VLSIDCS53788.2022.9811484.

Chapter 11

Sign language and hand gesture recognition using machine learning techniques

A comprehensive review

Janpreet Singh and Dalwinder Singh
Lovely Professional University, Phagwara, India

CONTENTS

DOI: 10.1201/9781003407409-11

11.1 INTRODUCTION

Gestures of hands are primarily utilized by human beings to communicate with others as well as to reveal their feelings and thoughts to other individuals [1]. It also assists in delivering significant information in daily routine interaction [2]. Additionally, a structured form of hand gestures is known as sign language, which consists of several signs as well as visual motions and is also utilized as a system that helps in interaction or communication [3, 4]. It is an effective and beneficial tool for those individuals who are speech impaired and deaf. The distinct body parts have been used in sign language such as body, fingers, head, hand, expressions of a face in order to convey the useful message during the communication [5]. It is also found that this language of the sign is not widely used by the deaf community, and also some of them are unable to perceive the correct message from it [6]. Hence, this is a major unsolved issue among deaf individuals, and as a result, they are incapable of interacting with others to deliver their message to the rest of the individuals.

It is also analyzed that the main body part that is considerably used during sign language is the upper body part [7]. The meaning of signs used in this language varies from location to location, and also, the shape of these signs is distinct for the same sentence [8]. There are a number of kinds of hand gestures such as communicative gestures, manipulative gestures, controlling gestures, and conversational gestures [9]. The considered sign language in this respective research work is a kind of communicative gesture [10]. This paper mainly concentrates on the recognition of sign language. Howbeit, as it is mentioned that sign language comes under the category of hand gesture, that is, communicative gesture, then it is essential to review the literature of recognition of hand gesture along with the literature of recognition of sign language.

The arrangement of this paper from this onwards has been made in such a manner that the various challenges and approaches related to the recognition of gestures are discussed in Section 11.1. After that, the several distinct techniques used by other researchers for the recognition of gestures on vision-based and sensor-based have been explained in Sections 11.2 and 11.3, respectively. In Section 11.2, the four major phases considered in the study, that is, data gathering, pre-processing of acquired data, segmentation, extraction of features, as well as classification, are also explained with literature work on the vision-based approach. Furthermore, Section 11.4 is all about a summary of the findings acquired from the literature review. At last, the ideas about future work and a conclusion are presented in Section 11.5 of this paper.

11.2 CHALLENGES IN RECOGNITION OF HAND GESTURES

Some complex and difficult processes are involved in the identification of gestures. These processes are analysis of motion, algorithms of machine learning, motion modeling, and identification of patterns [11]. The gesture recognition also includes some parameters, which are either non-manual or manual in nature [12]. The ability to predict a gesture is also influenced by the environment of structure, such as the brightness of the background and the movement's agility [13]. In the case of two-dimensional space, the variation in views also affects the appearance of a specific gesture. It is also found that in order to segment the hand gestures from a video or an image, the signer used to wear a glove with some color or a band [5, 14]. This technique will help in the segmentation phase to extract the required gesture effectively and easily [15, 16]. In other words, the complexity that can be faced in the process of segmentation during the identification of hand gestures will reduce or diminish.

Moreover, numerous issues appear in the case of recognition of gestures that are dynamic in nature, for instance, spatial complexity, repeatability, temporal variance, gesture region and orientation, as well as movement epenthesis [17]. To overcome these explained challenges, researchers are performing their commendable work by developing new and adequate models for the recognition of sign language. However, the performance is a single parameter by which one can evaluate whether a particular model or system is accurate or not. The performance of the developed system is also calculated by using some evaluation criteria such as user independence, real-time performance, robustness, and scalability of a specific system [18].

11.3 APPROACHES' TYPES

There are two types of approaches that are utilized by the researchers in order to achieve better results from hand gesture recognition, as shown in Figure 11.1.

The first approach of hand gesture recognition is the vision-based approach. In this approach, a video camera is used for acquiring the video or picture of gestures of hand [19]. Additionally, some required equipment are a single camera, which can be a camera of a smartphone, or a video camera or webcam, a stereo camera, which assists in offering thorough information and a monocular camera are mostly utilized [20, 21]. Moreover, active techniques are also required in vision-based approach that utilizes the light's projection, and at last, the technique of using lights, bands, or gloves to reduce complexity comes under invasive techniques [22].

The second approach of gesture recognition is a sensor-based approach. In this approach, the hand's velocity, position as well as motion are captured by using instruments, and also in this, sensors are majorly utilized [23]. IMU, that is, inertial measurement unit, is used to evaluate the finger's speed, degree of freedom, as well as position of the hand. It consists of the utilization of an accelerometer as well as a gyroscope. For the detection of movements of fingers as well as to determine the pulse of muscles of a human's

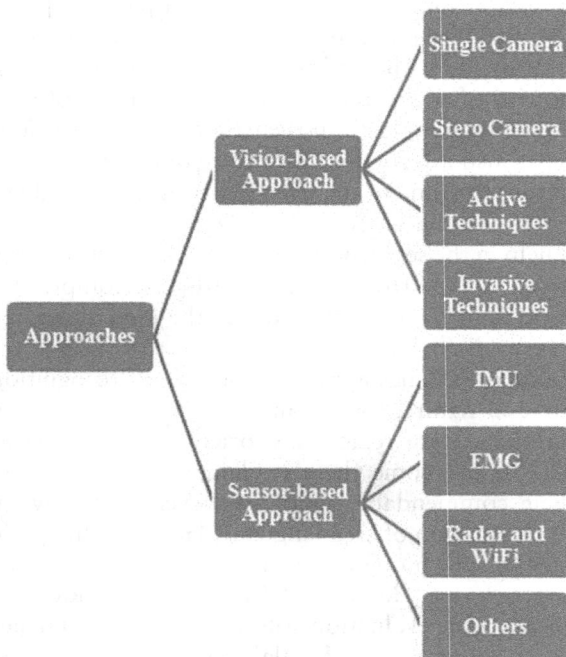

Figure 11.1 Types of approaches.

hand, EMG or electromyography (EMG) is employed [24]. Moreover, the change in in-air signal strength is detected by using a radar's beam. Haptic, mechanical, ultrasonic, as well as electromagnetic techniques, flex sensors are other instruments that are exploited in this type of approach [25].

11.4 LITERATURE REVIEW ON VISION-BASED GESTURE RECOGNITION

Figure 11.2 presents the number of phases that are involved in gesture recognition, that is, data gathering, pre-processing of acquired data, segmentation, extraction of features, as well as classification. The recognition of sign language or gestures can be static or dynamic [26]. The images or picture's frame are used in static gesture recognition and in contrast, continuous image frames, that is, video is used in case of identification of gestures, which are dynamic in nature [27]. The major difference between the two approaches of gesture recognition is seen in the methodology used for the acquisition of data [28]. This section entirely concentrates on the research work done by other authors on vision-based approach for the recognition of hand gestures.

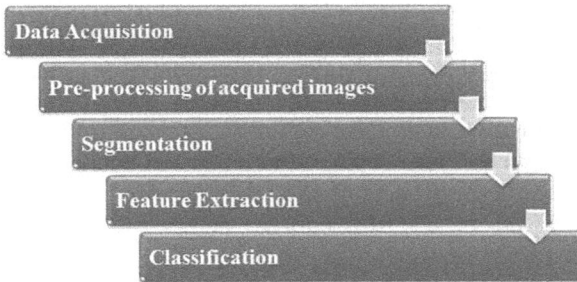

Figure 11.2 Phases of vision-based approach of gesture recognition.

11.5 DATA ACQUISITION

The image's frames are gathered as the data in vision-based approach of identification of gestures [29]. These frames are acquired by using various equipment as well as techniques. The equipment are a thermal camera, web cam, or standard video camera. Likewise, invasive as well as active techniques are used to acquire the images effectively and accurately. To grab the thorough data from the images, several tools such as LMC and Kinect are used [30, 31]. In this way, the images or frames of pictures being collected is the very first of sign or hand gesture recognition.

11.6 PRE-PROCESSING OF ACQUIRED DATA

In order to enhance the performance of the developed system or model and to make amendments in the acquired images or videos, the pre-processing phase is performed. Some filters are utilized by the researchers that assist in diminishing the noise from gathered data, and these filters are the Gaussian filter and Median filter [32]. Munasinghe [33] used the median filter to lower the noise present in the acquired frames in his research work. Similarly, in the research work of Kakkoth [34], the median filter has been used with morphological operators to remove the considerable noise without any loss of useful information from the acquired picture, such as boundary information. Moreover, Pansare, Gawande, and Ingle [35] also utilized both Gaussian and a median filter along with morphological operator in their research work in order to diminish the unwanted data in the phase of pre-processing.

The computational efficiency is also reduced in some research studies by lowering the resolutions of collected images during the pre-processing phase. For instance, N. H. Dardas and N. D. Georganas [36] used to reduce the image's resolution, which will be used as an input to the system. It is analyzed that this technique helps to decrease the time of image processing by diminishing the various key points of a particular picture. Likewise, Rokade, Doye and Kokare [37] and Jayaprakash and Majumder [38] utilized the same methodology to decrease the processing time of an image without any effect on the overall performance of the system.

Furthermore, histogram equalization also assists in improving the image contour captured in the distinct surrounding. It will help to make the illumination as well as the brightness of the taken picture in a uniform manner. Various studies [28, 30, 39, 40] reported on the use of this technique to increase the picture's contour.

11.7 SEGMENTATION

A process in which the captured pictures or images are partitioned into a number of different parts is known as segmentation [41]. The required part of an image, also known as region of interest or ROI, is partitioned in this respective process from the whole picture [42]. Additionally, the process of segmentation has two types. First is contextual segmentation, in which the association of spatial among the various features has been taken under consideration, for instance, the technique for the identification of edges. The second type of segmentation is non-contextual segmentation, in which these relationships are not taken into account; however, the group pixels are dependent upon the global features [43].

11.7.1 Segmentation of skin color

HIS, HSV, YCbCr, and RGB color spaced are majorly used in the segmentation of skin color [44]. In order to avail a robust segmentation of skin color, the researcher faced various challenges such as characteristics of the camera, illumination, sensitivity, as well as color of skin [45]. The palm and the hand can be segmented accurately and easily by using HSV, as this color space is famous for partitioning the hue of the hand as well as arm [46]. In the research work [13, 36], HSV was successfully used to partition the hand from the face during the skin color segmentation process. Similarly, the researchers who used RGB for skin color segmentation in their research work are [47, 48]. This color space is most popular as well as utilized in digital pictures and encodes three colors, which are primary in nature. R. F. Rahmat, T. Chairunnisa, D. Gunawan and O. S. Sitompul [49] and S. Kolkur, D. Kalbande, P. Shimpi, C. Bapat, and J. Jatakia [50] combined the three-color space with each other in their work to achieve an accurate and better result. These color spaces are RGB, YCbCr, as well as HSV. Similarly, these color spaces are also compared by K. Basha, P. Ganesan, V. Kalist, B. S. Sathish, and J. M. Mary [51]. According to this research work, YCbCr color space provides effective and adequate results in the case of segmentation and identification of color of skins.

11.7.2 Other methods for segmentation

S. M. Nadgeri, S. D. Sawarkar, and A. D. Gawande [52] developed a sign language recognition system by using CAMShift or Continuous Adaptive Mean Shift Algorithm. This algorithm helps in tracking the objects having colors in acquired data, that is, various sequences of frames of videos. Color histograms are also made in this research work to represent the color pictures as a probability distribution. There are three main phases of this research work, that is, detection of skin color, tracking of hand, as well as recognition of hand gestures.

R. F. Rahmat, T. Chairunnisa, D. Gunawan, and M. F. Pasha [53] provided a system for hand gesture recognition from various applications. These applications include video, music player, presentation, and PDF reader. The main challenge which was overcome in this work is different illumination conditions. This presented system aids to detect the gesture even with a complex background. The two color spaces are being combined in order to solve the issue of complex backgrounds in the dataset. The used color space is HS-CbCr and has 96.87% accurate results in good illumination.

T. Zhang, H. Lin, Z. Ju, and C. Yang [54] presented a hand gesture recognition system having two phases and helping to resolve the complex background problem. The convolutional pose machine is used to acquire the required key points from the data's complex backgrounds. The RGB picture is provided as an input to the convolutional pose machine, and as an output, the machine will generate a heatmap for every required point of the hand. In the second phase, the Fuzzy Gaussian mixture model is used to classify the hand gestures effectively.

Wu and Wang [55] also used a convolutional pose machine in their research work to acquire useful information from the gathered dataset. However, the classification method in this study is a convolutional neural network that aids to classify the provided hand gestures as well as sign languages into the correct class.

11.7.3 Tracking

Another crucial part of segmentation is tracking, as the segmentation as well as tracking are utilized to acquire the hand from images or videos having complex and complicated backgrounds. It is very challenging to track a hand from a video because the hand's appearance and movement are continuously as well as rapidly changing within few seconds. In some studies, the CAMShift has been utilized for tracking the movement of the hand [3, 54, 55]. This methodology assists in finding the hand gesture's location and also avoids the use of other techniques in order to detect the gestures' position.

Ding, Jiang, and Zou [32] used Adaboost to develop a system for dynamic gesture recognition. This paper also stated that this method provides more accuracy and has better performance when compared to other machine learning algorithms. The Adaptive Boosting algorithm is comprised of the linear combination of various weak classifiers. It can also be applied in order to detect three-dimensional gestures [56]. Hence, it is not inaccurate to state the problem of three-dimensional gesture recognition can be solved by employing the Adaptive Boosting algorithm.

11.8 FEATURE EXTRACTION

Feature extraction is a process in which the required or interesting features are collected from a set of features. As this research work is all about identification gestures, hence, the desired feature for this is the information about gestures of hand that are used as an input to the system.

11.8.1 Principal component analysis (PCA)

The principal component analysis (PCA) reduces the data dimensionality of the selected dataset. This results in the extraction of features that are required as well as ignoring the unnecessary information [57]. Assume a set of images A for the training process with an N-dimensional vector. The PCA evaluates a subspace having s-dimensions. It is also stated that the newly created subspaces' dimensions are always lower than already existed, which means that s << N. Equation 11.1 provides the mean of all the images that are considered during training, and it is represented by μ. Also, the $\underline{x}_j$ is the jth image.

$$\mu = \frac{1}{A} \tag{11.1}$$

Similarly, Equation 11.2 computed the scatter matrix S_T

$$S_T = \sum_{j=1}^{A} (x_j - \mu).(x_j - \mu)^2 \tag{11.2}$$

The eigenvalues, as well as eigenvectors, are evaluated, and after that, obtained eigenvectors are saved in their descending order. In order to reduce the dimensionality of a given dataset, the eigenvectors which are having lower values of eigenvalues are removed from the dataset, as these eigenvectors are not containing useful information [58].

Kuman, Srivastava, and Singh [59] stated that the PCA is a robust and effective tool for analyzing the data. They used PCA in their research work for dimensional reduction and extraction of desired features. It is also analyzed that the PCA is a tool that provides better results and even the considered data has noise in it. The accuracy of the developed model is observed as 100%. Due to the useful characteristics of PCA, Alejo and Funes [60]also used it in their work for the extraction of features and to reduce the noise from the collected dataset. The results have been stated that the proposed system has 94.74% accuracy for recognizing the hand gesture from a video. The PCA method also provides a roadmap to S. R. Kota, J. L. Raheja, A. Gupta, A. Rathi, and S. Sharma [60] about how the complicated dataset having a number of dimensions can be reduced. A system has been developed to identify the gestures of humans from two-dimensional space by using PCA and Euclidean distance.

11.8.2 Linear discriminant analysis (LDA)

Linear discriminant analysis or LDA is an approach that is utilized to figure out the features' linear combinations, which separate two or more data's classes. It is expected that the classes are normally distributed in LDA. Equation 11.3 provides the formulae for S_w and S_B representing scatter matrices of within-class and between-class, respectively.

$$S_W = \sum_{j=1}^{N} \sum_{x_k \in X_j} (x_k - \mu_j).(x_k - \mu_j)^T$$

$$S_B = \sum_{j=1}^{N} N_j (x_j - \mu).(x_j - \mu)^T \tag{11.3}$$

where N_j is the number of samples used in the training process in j class, the mean vector of samples is represented by μ, and x_k is class' kth image.

Jasim and M. Hasanuzzaman [61] used two methods for the extraction of features separately. These two approaches are Local Binary Pattern and Linear Discriminant Analysis. The Bangladeshi, as well as the Chinese language, has been recognized in this work, and the nearest neighbor algorithm is employed for the classification of gestures. Due to the implementation of LDA, the observed accuracy is 88.55% and 92.417% for Bangladeshi and Chinese languages, respectively. The combination of support vector machine (SVM) and linear discriminant analysis has been used by Z. Li [62] for feature extraction and classification of recognized gestures. The recognition rate of this model with six electrodes is 97.61%. Similar models have been developed by researchers [63, 64] for the recognition of hand gestures and sign language but with different classifiers.

11.8.3 Shift-invariant feature transform (SIFT)

Shift-invariant feature transform (SIFT) is presented by Lowe, and it is a technique that assists in the extraction of the feature that is rotation as well as scale-invariant [65]. It explains a selected picture according to only the points, which are interesting or required. The Gaussian Function is implemented in each phase of SIFT in order to smooth as well as rescale an image. Equation 11.4 shows the function by which scale spaced is described.

$$L(x,y,\sigma) = G(x,y,\sigma) \times I(x,y) \tag{11.4}$$

The difference of Gaussian function (DoG) is used to calculate the minima as well as maxima of extracted key points of an image. The number of applications of gesture identification employed SIFT to extract the desired and useful features from the selected dataset. B. Gupta, P. Shukla, and A. Mittal [66] develop an approach for the identification of sign language from static images. The feature extraction has been done by implementing HOG and SIFT. The KNN classifier is also utilized in order to classify the recognized gestures with an accuracy of 84.23%. The research work done by Benmoussa and Mahmoudi [67] acquired 98% performance for the recognition of 16 different gestures by using SURF and SIFT descriptors. This system used a SVM for the classification of gestures. An identical system has been presented by H. Mahmud, M. K. Hasan, A. A. Tariq, and M. A. Mottalib [68], and it successfully achieved 99 percentile accuracy. Pandita and Narote [69] concluded that SIFT is an effective approach for feature extraction. The extracted features by this methodology are interesting, as these features have a distinctiveness, which further helps to match the recognized features with a correct and accurate class of gestures.

11.9 CLASSIFICATION

The supervised as well as unsupervised machine learning approaches are implemented for the classification. A supervised approach is that in which the training is provided to the system by using various input data, which further assist the system to predict any data given to it [70]. In contrast, in the case of an unsupervised approach, the system will gain knowledge from its experience, and the accuracy of the system automatically increases as it predicts the number of patterns by itself.

11.9.1 Artificial neural network (ANN)

The working of an artificial neural network (ANN) is identical to the working of biological neural networks [71]. It handles several complex computations [72, 73]. Figure 11.3 provides the block diagram of ANN.

S. K. Yewale and P. K. Bharne [75] used MATLAB software to develop an interface with ANN in order to recognize hand gestures. In the research work of Stergiopoulou and Papamarkos [76], a similar concept has been

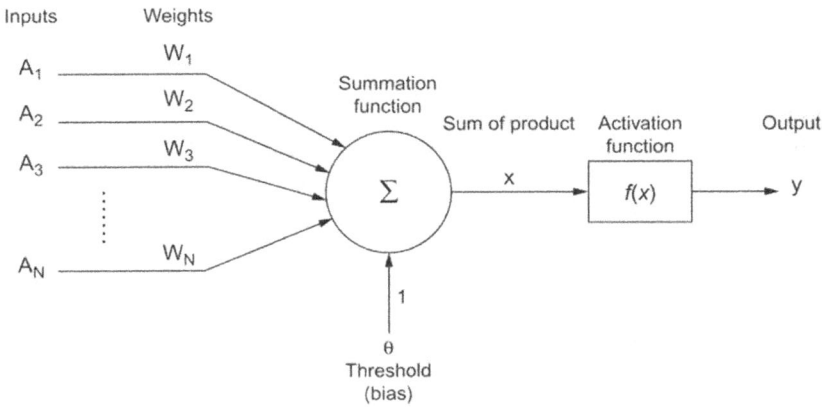

Figure 11.3 Block diagram of ANN [74].

implemented and successfully achieved a 90.45% of recognition rate. One thousand images from the database were utilized by Ekbote [4] for the identification of sign language, and after that, ANN as well as SVM classifier are used for their accurate classifications. The various gestures can be converted into useful sentences by using ANN, which helps to become interaction easier between normal and disabled individuals [77].

11.9.2 Support vector machine (SVM)

Support vector machine is a ML approach having labels, and the main intention of this approach is to identify a hyperplane that is optimal and separates the several datapoints [78]. The hand gestures were identified by using SVM from the video clips by [42]. The accuracy for detecting a word by using this system is 94.26%. Chen and Zhang [79] also proposed a system with a recognition rate of 89% and utilized the SVM approach with surface EMG in order to identify the hand gestures.

11.9.3 Euclidean distance classifier

Tripathi, Baranwal, and Nandi [80] introduced a new system to identify the Indian sign language. This research work used Manhattan distance, Euclidean Distance, and correlation for the classification and obtained better results. Similarly, Singha and Das [81] described the approach presented by them for the detection of sign language. It also used Euclidean distance and achieved a 96 percentile recognition rate. The same classifier is also utilized by these researchers [82] and hits a 96.25% recognition rate this time. Kagalkar and Gumaste [83] identified the Kannada language by using Euclidean distance from the videos. These videos comprised of a number of signs, and after performing various phases, this work acquired a 95.25% hit rate.

11.9.4 K-nearest neighbor classifier

In the case of KNN, the given data is classified into its accurate class according to the majority votes of its neighbor. It is an effective as well as a simple approach for the classification of any input data [84]. The combination of Euclidean distance and KNN is employed in a system so that the particular system is able to identify the sign language of India. However, this developed system is not accurate for all the gestures [85]. Yadev [86] also presented a system that can be used for the identification of sign language. In this research work, the KNN classifier is used for classification purposes, and the gestures are recognized by using the dataset of images. The model provided by Nogales and Benalcázar [87] assisted to recognize static gestures and used the KNN algorithm for classification. This model achieved 92.22% accuracy.

11.10 LITERATURE REVIEW ON SENSOR-BASED GESTURE RECOGNITION

The various techniques that are utilized in the researcher's work for the recognition of sensor-based gestures are discussed in this section. In this approach, the sensors are connected with the user physically in order to acquire information about hands and finger's trajectories, motions, as well as position. The use of this approach in gesture recognition assists in removing the requirement of segmentation as well as pre-processing phase, whereas these phases are most crucial in the case of vision-based approach. In this respective approach, one mainly needs to wear a glove or a band that must be connected to the user's arm.

11.10.1 Data glove

In order to retrieve the information about acceleration, angular, as well as orientation of gesture, the IMU sensors such as accelerometer and gyroscope are utilized by the data gloves to recognize the sign language or a gesture. The sensors which are used to access the data regarding bending of the fingers are known as flex sensors. A simple model for the recognition of sign language has been developed by using data gloves in research [88]. The proposed methodology in this paper is able to detect lip motions, expressions of the face, head gestures, as well as hand gestures. In a study by A. Z. Shukor [89], the Malaysian sign language has been detected by using the proposed system. In this research, the data gloves' configuration consists of 10 tilt sensors, which further help to acquire the motion of hand and flexion of fingers. The experimental result of this work shows that the tilt sensors must be inclined to more than 85 degrees to gain better results. This system provides

95 percentile accuracy for translating alphabets, 78.33 percentile for gestures, and 93.33 percentile accuracy for numbers. P. Lokhande, R. Prajapati, and S. Pansare [90] also presented an effective methodology that assists deaf and dumb individuals to interact with others. The researchers provided an embedded system that further recognizes sign language. The data glove is used in this system, which has to be worn by the dumb individual in order to interact with normal individuals. Three-axis accelerometers, as well as flex sensors, are also used to transform the provided gestures of hand into the text. A review is also provided by M. A. Ahmed, B. B. Zaidan, A. A. Zaidan, M. M. Salih, and M. M. Lakulu [91] regarding those research works in which data gloves have been used to recognize the sign language from 2007 to 2017. This paper constructed an effective roadmap for the readers to figure out the research gap as well as the research work that has been done by other researchers so far.

11.10.2 Electromyography (EMG)

The recording of the actions, which are electrical in nature and are performed by muscle tissues by utilizing electrodes embedded either into muscles or to the skin, is known as EMG [92]. E. Ceolini, C. Frenkel, S. B. Shrestha, G.Taverni, L. Khacef, M. Payvand, and E. Donati [93] presented a framework that is sensor fusion and merges complementary systems such as signals of EMG from an individual's muscle and visual data. The recording of these signals has been done by using conventional electrodes. Although this presented multisensory methodology assists to enhance the robustness as well as the accuracy of the developed system but also the computational cost of embedding sensors is exponentially increased. Jaramillo and Benalcazer [94] proposed an approach that can be used for the recognition of hand gestures in real-time applications. The following approach uses EMG signals as well as a machine learning algorithm for the classification of gestures. The accuracy of this system is observed up to 86 percentile. Similarly, M. E. Benalcazar, J. González, A. Jaramillo-Yánez, C. E. Anchundia, P. Zambrano, and M. Segura [95] presented a model, which can detect five different gestures such as double tap, wave out, wave in. finger spread, and fist. The 25 different repetitions of these mentioned gestures have been used in order to train the model. According to the experimental results, it is observed that the average processing time of this model is 40.58 ± 1.62 ms, and the accuracy is evaluated as 92.45% with a standard deviation of 11 percentile.

11.10.3 Radar and WiFi

WiFi-oriented gesture control is another type of approach that can be utilized for the detection of gestures. Many researchers stated that this technology is easier to implement in contrast to Kinect technology. This approach mainly concentrates on a single user in order to identify the various gestures of that particular user, as it consists of multiple antennas. The WiFi signals can pass through walls; hence, it does not need any kind of line of sight. This kind of work is done by Q. Pu, S. Gupta, S. Gollakota, and S. Patel [96]. Additionally, W. He, K. Wu, Y. Zou, and Z. Ming [97] gave WiG device for the identification of sign language. It consists of WiFi devices as well as infrastructure. This device has high practicability as well as very low cost. This device can also be employed in the indoor surrounding. The accuracy of this model is being evaluated on the basis of two scenarios, that is, non-line of sight and line of sight, and the results showed 88% and 92% accuracy, respectively. A model for gesture detection is also proposed by H. Abdelnasser, M. Youssef, and K. A. Harras [98] by utilizing the WiFi, and the name of this model is WiGest. It uses the signals of WiFi in order to recognize the gestures of hands of a specific user. By using one AP, the accuracy of this system is 87.5%, and moreover, this accuracy increases when 3 overhead access points are utilized to 96 percentile. Furthermore, a WiGAN is also developed by Jiang, Li, and Xu [99] in their research work. The Generative Adversarial Network has been used in this work in order to generate as well as extract the required features of gestures. For the classification of provided gestures, a SVM is utilized. The performance of the system has been evaluated by using two different CSI-based datasets and the accuracy of this model is 98 percentile and 95.6 percentile, respectively.

11.11 DISCUSSION

This section includes the critical review of surveys that are already done by other researchers on gesture and sign language recognition. Additionally, various techniques that are employed by the researchers in their corresponding work are also discussed.

11.11.1 Critical review on the previous survey

Several surveys, as well as comprehensive reviews, are provided by several researchers on the recognition of sign language and gestures. These surveys offer a detailed explanation of various approaches that are utilized for the identification of sign language as well as gestures. The analysis and main intend of these review papers are illustrated in Table 11.1.

11.11.2 Reviews on considered algorithms and methodologies

This section includes a summary of considered research works for sensor-based as well as vision-based gesture recognition. Tables 11.2 and 11.3 list

Table 11.1 Comprehensive reviews on the recognition of gestures and sign language

References	Author	Year	Focus
[10]	Wu and Huang	2001	The temporal, as well as static hand postures and the methods used for their identification, is being elaborated in this paper
[45]	Kakumanu et al.	2007	Reviewed several approaches used for the detection of skin and its color
[47]	Khattab et al.	2014	Compared distinct color spaces such as YUV, XYZ, CMY, HSV, and RGB and stated that RGB is an accurate and best color space to achieve the highest recognition rate.
[19]	Kumar and Saerbeck	2015	Discussed vision-based gesture recognition approaches from the previous 16 years
[51]	Shaik et al.	2015	Compared two color spaces, that is, YCbCr and HSV, for the segmentation and detection of skin color
[14]	Anderson et al.	2017	Discussed various techniques used for recognition of sign language and figured out which one is best
[6]	Cheok, Omar and Jaward	2017	Described the latest used methodologies for the recognition of SLR and for the recognition of hand gestures with their limitations and challenges
[91]	Ahmed et al.	2018	Explained the recent research work done in regards to SLR based on sensor gloves from 2007 to 2017

the number of research works that were taken under consideration in this work along with the methodologies and approaches used in that particular work for the recognition of sign language and hand gesture based on vision and sensor-based approaches. The phase of pre-processing is not included in this summary part. However, the methodologies used for feature extraction, classification, and segmentation in each research work are effectively summarized in both tables. A column is provided to the achieved sample size/recognition rate/accuracy by the proposed models. Here, sample size represented the sample used in particular work for both phases, that is, for testing as well as training. The scope of every research work is also summarized.

Table 11.2 Summary of vision-based gesture recognition approach

Reference	Author name	Year	Classification	Feature extraction	Segmentation	Sample size/ recognition rate/accuracy	Scope
[64]	Hongo et al.	2000	LDA	LDA	Skin color (LUV)	100% and 96.9%	Hand and face gestures
[44]	Lee et al.	2004	—	Entropy	Tracking	95%	Gestures
[53]	Rehmat et al.	2005	Count detect finger	Contour and Convex hull	Skin color (HS-CbCr)	96.87%	Gestures
[60]	Kota et al.	2009	Euclidean Distance	PCA	—	—	Gestures
[52]	Nadgeri et al.	2010	Euclidean Distance	Orientation Histograms	CAMShift	—	ASL
[76]	Stergiopoulou and Papamarkos	2011	ANN	Histogram	Tracking	90.45%	Hand gestures
[41]	Barkoky and Charkari	2011	SVM	Thinning method	Skin color information	96.62%	Persian sign numbers
[36]	Dardas and Georganas	2011	SVM	SIFT	Skin color (HSV)	96.23%	Gestures
[3]	Madani and Nahvi	2013	KNN, SVM, NN, MD	Radon transform and DCT	Tacking	95.56%	Persian sign language
[13]	Konwar et al.	2014	ANN	Edge detection	Skin color (HSV)	65%	ASL
[61]	Jasim and Hasanuzzaman	2014	Nearest Neighbor algorithm	LDA and local binary pattern	—	92.417% and 88.55%	Bangladeshi and Chinese numerals
[63]	Lee et al.	2015	Random Forest	LDA	Skin color	91.00%	Hand gestures

Ref.	Author	Year	Classifier	Feature extraction	Segmentation	Accuracy	Application
[49]	Rehmat et al.	2016	—	—	Skin color (RGB,YCbCr and HSV)	91.05%	Gestures
[62]	Li	2016	SVM	LDA	Skin color	97.61%	Gestures
[66]	Gupta et al.	2016	KNN	SIFT	Skin color	84.23%	Indian sign language gestures
[68]	Mahmud et al.	2016	SVM	SIFT	Skin color	640 * 480	Bangladeshi hand gestures
[42]	Santa et al.	2017	SVM	Local Binary Pattern	Skin color (RGB)	94.26%	
[50]	Kolkur et al.	2017	—	—	Skin color (RGB,YCbCr and HSV)	80.55%	Skin detection
[32]	Ding, Jiang and Zou	2017	Adaboost	PCA	Tracking	95.20%	Dynamic gestures
[59]	Kuman, Srivastava and Singh	2017	Euclidean Distance	PCA	Tracking	100%	Hand gestures
[43]	Huang et al.	2018	CNN	3D-CNN	Skin color (RGB)	82.7%	SLR
[55]	Wu and Wang	2018	CNN	Convolutional Pose Machine	Skin Color (RGB)	95.8%	Gestures
[67]	Benmoussa and Mahmoudi	2018	SVM	SIFT	Tracking	98%	Gestures
[100]	Alejo and Funes	2019	Euclidean Distance	PCA	Skin color (HSV)	94.74%	Dynamic gestures
[54]	Zang et al.	2020	Fuzzy Gaussian Mixture Model	Convolutional Pose Machine	Skin color (YUV, YCbCr)	98.06%	Hand gestures

Table 11.3 Summary of sensor-based gesture recognition approach

Reference	Author name	Year	Classification	Feature extraction	Sensor type	Sample size/ recognition rate/accuracy	Scope
[96]	Pu et al.	1999	Doppler Shifts	Extract features by evaluating signal's frequency time	WiSee	94%	Gestures
[88]	Jiangqin et al.	2002	Hidden Markov Model	Flex sensors	Data glove	90%	Chinese sign language
[89]	Shukor et al.	2015	Accelerometer	Flex sensors	Data glove	93.33%	Malaysian sign language
[90]	Lokhande et al.	2015	—	Flex sensor	Data glove	99%	Sign Language
[97]	He et al.	2015	SVM	Means, Standard deviation, median and maxima of patterns	WiG	92%	Gestures
[98]	Abdelnasser et al.	2015	Random Forest	Discrete Wavelet Transform	WiGest	96%	Gestures
[94]	Jaramillo and Benalcazer	2017	Naïve Bayes classifier, SVM and ANN	EMG signals	EMG	86%	Hand gestures
[93]	Coelini et al.	2020	Neural Network	Mean Absolute value	EMG	30*600	Hand gestures
[95]	Benalcazer et al.	2020	Feed forward ANN	EMG signals	EMG	92.85%	Hand gestures
[99]	Jiang, Li and Xu	2020	SVM	Generative Adversarial Network	WiGAN	98%	Gestures
[92]	Li, Shi and Yu	2021	Deep Learning, CNN	sEMG images	EMG	—	Gestures

11.12 CONCLUSION AND FUTURE WORK

There is a huge scope of gesture recognition in the research domain due to its immense potential in various applications. Applications include human-computer interaction, remote control robots, and recognition of sign languages. According to the reviewed literature related to the recognition of sign language and hand gestures, it is analyzed that there are several emerging methodologies that are gaining interest in this domain. The technologies that can be used to acquire the information about provided gestures easily and effectively are Kinect and EMG. The pre-processing of the considered dataset can be done more appropriately and adequately by employing Gaussian and Median filters. For feature extraction, SIFT is considered as the best approach, which assists in achieving the highest results. Among various color spaces, HSV and YCbCr are two majorly utilized color spaced for skin color segmentation. Similarly, a SVM is also observed as a growing approach that can used in order to classify the recognized gestures into a correct class.

REFERENCES

1. P. Gairola, and S. Kumar, "Hand gesture recognition from video," *International Journal of Science and Research (IJSR)*, vol. 3, no. 4, pp. 154–158, 2014.
2. C. M. Jin, Z. Omar, and M. H. Jaward, "A mobile application of American sign language translation via image processing algorithms," in *2016 IEEE Region 10 Symposium (TENSYMP)*, 2016.
3. J. Singh, H. Singh, and D. K. Singh, "A novel approach based on gesture recognition through video capturing for sign language," in *2019 2nd International Conference on Intelligent Computing, Instrumentation and Control Technologies (ICICICT)* (Vol. 1, pp. 1177–1181). IEEE, July 2019.
4. J. Singh, S. Harjeet, and V. Goyal, "A comprehensive study on feature extraction and classification techniques for sign language recognition," *2021 5th International Conference on Electronics, Communication and Aerospace Technology (ICECA)*. IEEE, 2021.
5. P. Shukla, A. Garg, K. Sharma, and A. Mittal, "A DTW and Fourier Descriptor based approach for Indian sign language recognition," in *2015 Third International Conference on Image Information Processing (ICIIP)*, 2015.
6. M. Jin, C. Zaid, O. Mohamed, and H. Jaward, "A review of hand gesture and sign language recognition techniques," *International Journal of Machine Learning and Cybernetics*, vol. 1, pp. 23–41, 2017.
7. U. Bellugi, and S. Fischer, "A comparison of sign language and spoken language," *Cognition*, vol. 1, no. 2–3, pp. 173–200, 1972.
8. R. Yang, S. Sarkar, and B. Loeding, "Handling movement epenthesis and hand segmentation ambiguities in continuous sign language recognition using nested dynamic programming," *IEEE Transactions on Pattern Analysis and Machine Intelligence*, vol. 32, no. 3, pp. 462–477, 2010.

9. Y. Wu, and T. S. Huang, "Human hand modeling, analysis and animation in the context of HCI," in *Proceedings 1999 International Conference on Image Processing (Cat. 99CH36348)*, 2003.

10. Y. Wu, and T. S. Huang, "Vision-based gesture recognition: A review," *Gesture-Based Communication in Human-Computer Interaction*, vol. 1, pp. 103–115, 1999.

11. B. Mocialov, G. Turner, K. Lohan, and H. Hastie, "Towards continuous sign language recognition with deep learning," 2017.

12. K. Grobel, and M. Assan, "Isolated sign language recognition using hidden Markov models," in *1997 IEEE International Conference on Systems, Man, and Cybernetics. Computational Cybernetics and Simulation*, 2002.

13. A. S. Konwar, B. S. Borah, and C. T. Tuithung, "An American sign language detection system using HSV color model and edge detection," in *2014 International Conference on Communication and Signal Processing*, 2014.

14. R. Suharjito, F. Anderson, M. Wiryana, C. Ariesta, and G. P. Kusuma, "Sign language recognition application systems for deaf-mute people: A review based on input-process-output," *Procedia Computer Science*, vol. 116, pp. 441–448, 2017.

15. T. E. Starner, MIT Department of Brain and Cognitive Sciences, 1995.

16. "4 year project report hand gesture recognition using computer," yumpu.com. [Online]. Available: https://www.yumpu.com/en/document/view/5329170/4-year-project-report-hand-gesture-recognition-using-computer. [Accessed: 28-Dec-2021].

17. Y. Nam, and K. Wohn, "Recognition of space-time hand-gestures using hidden Markov model," in *Proceedings of the ACM Symposium on Virtual Reality Software and Technology - VRST '96*, 1996.

18. K. Arora, A. Kumar, V. K. Kamboj, D. Prashar, and B. Shrestha, G.P. Joshi, "Impact of renewable energy sources into multi area multi-source load frequency control of interrelated power system," *Mathematics*, vol. 9, p. 186, 2021.

19. P. K. Pisharady, and M. Saerbeck, "Recent methods and databases in vision-based hand gesture recognition: A review," *Computer Vision and Image Understanding*, vol. 141, pp. 152–165, 2015.

20. H. Cooper, B. Holt, and R. Bowden, "Sign language recognition Visual Analysis of Humans," pp. 539–562, 2011.

21. P. V. V. Kishore, A. S. C. S. Sastry, and A. Kartheek, "Visual-verbal machine interpreter for sign language recognition under versatile video backgrounds," in *2014 First International Conference on Networks & Soft Computing (ICNSC2014)*, 2014.

22. L. Geng, X. Ma, B. Xue, H. Wu, J. Gu, and Y. Li, "Combining features for Chinese sign language recognition with Kinect," in *11th IEEE International Conference on Control & Automation (ICCA)*, 2014.

23. C. Keskin, F. Kıraç, Y. E. Kara, and L. Akarun, "Real time hand pose estimation using depth sensors," in *Consumer Depth Cameras for Computer Vision*, London: Springer, 2013, pp. 119–137.

24. A. R. Asif et al., "Performance evaluation of convolutional neural network for hand gesture recognition using EMG," *Sensors (Basel)*, vol. 20, no. 6, p. 1642, 2020.

25. J. L. Raheja, A. Mishra, and A. Chaudhary, "Indian sign language recognition using SVM," *Pattern Recognition and Image Analysis*, vol. 26, no. 2, pp. 434–441, 2016.

26. R. Hartanto, A. Susanto, and P. I. Santosa, "Preliminary design of static indonesian sign language recognition system," in *2013 International Conference on Information Technology and Electrical Engineering (ICITEE)*, 2013.
27. A. Ghotkar, and G. Kharate, "Dynamic hand gesture recognition for sign words and novel sentence interpretation algorithm for Indian sign language using Microsoft Kinect sensor," *Journal of Pattern Recognition Research*, vol. 10, no. 1, pp. 24–38, 2015.
28. E. Escobedo, and G. Camara, "A new approach for dynamic gesture recognition using skeleton trajectory representation and histograms of cumulative magnitudes," in *2016 29th SIBGRAPI Conference on Graphics, Patterns and Images (SIBGRAPI)*, 2016.
29. S.B. Carneiro, E.D.D.M. Santos, M.D.A. Talles, J.O. Ferreira, S.G.S. Alcalá, and A.F. Da Rocha, "Static gestures recognition for Brazilian sign language with kinect sensor," in *2016 IEEE Sensors*, 2016.
30. C. Chansri, and J. Srinonchat, "Reliability and accuracy of Thai sign language recognition with kinect sensor," in *2016 13th International Conference on Electrical Engineering/Electronics, Computer, Telecommunications and Information Technology (ECTI-CON)*, 2016.
31. S. Aliyu, M. Mohandes, M. Deriche, and S. Badran, "Arabie sign language recognition using the Microsoft Kinect," in *2016 13th International Multi-Conference on Systems, Signals & Devices (SSD)*, 2016.
32. X. Ding, T. Jiang, and W. Zou, "A new method of dynamic gesture recognition using Wi-Fi signals based on Adaboost," in *2017 17th International Symposium on Communications and Information Technologies (ISCIT)*, 2017.
33. M. I. N. P. Munasinghe, "Dynamic hand gesture recognition using computer vision and neural networks," in *2018 3rd International Conference for Convergence in Technology (I2CT)*, 2018.
34. K. Arora, A. Kumar, V. K. Kamboj, D. Prashar, S. Jha, B. Shrestha, G.P. Joshi, "Optimization methodologies and testing on standard benchmark functions of load frequency control for interconnected multi area power system in smart grids," *Mathematics*, vol. 8, p. 980, 2020.
35. J. R. Pansare, S. H. Gawande, and M. Ingle, "Real-time static hand gesture recognition for American sign language (ASL) in complex background," *Journal of Signal and Information Processing*, vol. 03, no. 03, pp. 364–367, 2012.
36. N. H. Dardas, and N. D. Georganas, "Real-time hand gesture detection and recognition using bag-of-features and support vector machine techniques," *IEEE Transactions on Instrumentation and Measurement*, vol. 60, no. 11, pp. 3592–3607, 2011.
37. R. Rokade, D. Doye, and M. Kokare, "Hand gesture recognition by thinning method," in *2009 International Conference on Digital Image Processing*, 2009.
38. R. Jayaprakash, and S. Majumder, "Hand gesture recognition for sign language: A new hybrid approach," *International Conference on Image Processing*, vol. 1, pp 1981–1993, 2011.
39. J. R. Pansare, and M. Ingle, "Vision-based approach for American sign language recognition using Edge Orientation Histogram," in *2016 International Conference on Image, Vision and Computing (ICIVC)*, 2016.
40. A. Sethi, K. Kumar, and B. Rao, "SignPro-An application suite for deaf and dumb," *IJCSET*, vol. 2, no. 5, pp. 1203–1206, 2012.

41. A. Barkoky, and N. M. Charkari, "Static hand gesture recognition of Persian sign numbers using thinning method," in *2011 International Conference on Multimedia Technology*, 2011.

42. U. Santa, F. Tazreen, and S. A. Chowdhury, "Bangladeshi hand sign language recognition from video," in *2017 20th International Conference of Computer and Information Technology (ICCIT)*, 2017.

43. J. Huang, W. Zhou, Q. Zhang, H. Li, and W. Li, "Video-based sign languagerecognition without temporal segmentation," arXiv [cs.CV], 2018.

44. J. Lee, Y. Lee, E. Lee, and S. Hong, "Hand region extraction and gesture recognition from video stream with complex background through entropy analysis," *Conference Proceedings of the IEEE Engineering in Medicine and Biology Society*, vol. 2004, pp. 1513–1516, 2004.

45. P. Kakumanu, S. Makrogiannis, and N. Bourbakis, "A survey of skin-color modeling and detection methods," *Pattern Recognition*, vol. 40, no. 3, pp. 1106–1122, 2007.

46. L. Yun, Z. Lifeng, and Z. Shujun, "A hand gesture recognition method based on multi-feature fusion and template matching," *Procedia Engineering*, vol. 29, pp. 1678–1684, 2012.

47. D. Khattab, H. M. Ebied, A. S. Hussein, and M. F. Tolba, "Color image segmentation based on different color space models using automatic GrabCut," *Scientific World Journal*, vol. 2014, p. 126025, 2014.

48. H. K. Saini, and O. Chand, "Skin segmentation using RGB color model and implementation of switching conditions," *International Journal of Engineering Research and Applications (IJERA)*, vol. 3, no. 1, pp. 1781–1787, 2013.

49. R. F. Rahmat, T. Chairunnisa, D. Gunawan, and O. S. Sitompul, "Skin color segmentation using multi-color space threshold," in *2016 3rd International Conference on Computer and Information Sciences (ICCOINS)*, 2016.

50. S. Kolkur, D. Kalbande, P. Shimpi, C. Bapat, and J. Jatakia, "Human skin detection using RGB, HSV and YCbCr color models," in *Proceedings of the International Conference on Communication and Signal Processing 2016 (ICCASP 2016)*, 2017.

51. K. Basha, P. Ganesan, V. Kalist, B. S. Sathish, and J. M. Mary, "Comparative study of skin color detection and segmentation in HSV and YCbCr Color Space," *Procedia - Procedia Computer Science*, vol. 57, pp. 41–48, 2015.

52. S. M. Nadgeri, S. D. Sawarkar, and A. D. Gawande, "Hand gesture recognition using CAMSHIFT algorithm," in *2010 3rd International Conference on Emerging Trends in Engineering and Technology*, 2010.

53. R. F. Rahmat, T. Chairunnisa, D. Gunawan, and M. F. Pasha, "Hand gestures recognition with improved skin color segmentation in human-computer interaction applications," *Journal of Theoretical and Applied Information Technology*, vol. 97, no. 3, pp. 727–739, 2019.

54. T. Zhang, H. Lin, Z. Ju, and C. Yang, "Hand gesture recognition in complex background based on convolutional pose machine and Fuzzy Gaussian mixture models," *International Journal of Fuzzy Systems*, vol. 22, no. 4, pp. 1330–1341, 2020.

55. Y. Wu, and C. -M. Wang, "Applying hand gesture recognition and joint tracking to a TV controller using CNN and Convolutional Pose Machine," in *2018 24th International Conference on Pattern Recognition (ICPR)*, 2018.

56. J. Sheng, *A Study of Adaboost in 3D Gesture Recognition*, USA: Department of Computer Science, University of Toronto, 2003.
57. H. Nikita, and B. Sadawarti, "Classification of renal cancer using principal component analysis (PCA) and K-nearest neighbour (KNN)," *International Journal of Engineering Research & Technology (IJERT)*, vol. 8, no. 16, pp. 34–42, 2020.
58. W. Hai, "Gesture recognition using principal component analysis, multi-scale theory, and hidden Markov models: Semantic scholar," *Semantic Scholar*, 01-Jan-1970. [Online]. Available: https://www.semanticscholar.org/paper/Gesture-recognition-using-principal-component-and-Hai/3b6346ec4cca5edc83b2235a7c77caa78088e1bd. [Accessed: 25-Sep-2021].
59. S. Kumar, T. Srivastava, and R. S. Singh, "Hand gesture recognition using principal component analysis," *Asian Journal of Engineering and Technology*, vol. 6, pp. 32–35, 2017.
60. S. R. Kota, J. L. Raheja, A. Gupta, A. Rathi, and S. Sharma, "Principal component analysis for gesture recognition using SystemC," in *2009 International Conference on Advances in Recent Technologies in Communication and Computing*, pp. 732–737, 2009, doi: 10.1109/ARTCom.2009.177.
61. M. Jasim, and M. Hasanuzzaman, "Sign language interpretation using linear discriminant analysis and local binary patterns," in *2014 International Conference on Informatics, Electronics & Vision (ICIEV)*, pp. 1–5, 2014, doi: 10.1109/ICIEV.2014.7136001.
62. Z. Li, "A fast detection and recognition algorithm with construction of fast support vector machine based on entropy Weight," *International Journal of Security and Its Applications*, vol. 10, no. 8, pp. 23–28, 2016.
63. O. Sangjun, R. Mallipeddi, and M. Lee, "Real time hand gesture recognition using random forest and linear discriminant analysis," in *Proceedings of the 3rd International Conference on Human-Agent Interaction*, 2015.
64. H. Hongo, M. Ohya, M. Yasumoto, and K. Yamamoto, "Face and hand gesture recognition for human-computer interaction," in *Proceedings 15th International Conference on Pattern Recognition (ICPR-2000)*, vol. 2, pp. 921–924, 2000, doi: 10.1109/ICPR.2000.906224.
65. D. G. Lowe, "Distinctive image features from Scale-invariant keypoints," *International Journal of Computer Vision*, vol. 60, no. 2, pp. 91–110, 2004.
66. B. Gupta, P. Shukla, and A. Mittal, "K-nearest correlated neighbor classification for Indian sign language gesture recognition using feature fusion," in *2016 International Conference on Computer Communication and Informatics (ICCCI)*, pp. 1–5, 2016, doi: 10.1109/ICCCI.2016.7479951.
67. M. Benmoussa, and A. Mahmoudi, "Machine learning for hand gesture recognition using bag-of-words," in *2018 International Conference on Intelligent Systems and Computer Vision (ISCV)*, pp. 1–7, 2018, doi: 10.1109/ISACV.2018.8354082.
68. H. Mahmud, M. K. Hasan, A. A. Tariq, and M. A. Mottalib, "Hand gesture recognition using SIFT features on depth image," in *The Ninth International Conference on Advances in Computer-Human Interactions*, 2016.
69. S. Pandita, and S. P. Narote, "Hand gesture recognition using SIFT," *International Journal of Engineering Research & Technology (IJERT)*, vol. 2, no. 1, pp. 2–5, 2013.

70. S. Kavitha, S. Varuna, and R. Ramya, "A comparative analysis on linear regression and support vector regression," *2016 Online International Conference on Green Engineering and Technologies (IC-GET)*, pp. 1–5, 2016, doi: 10.1109/GET.2016.7916627.

71. S. E. Dreyfus, "Artificial neural networks, back propagation, and the Kelley-Bryson gradient procedure,"*Journal of Guidance, Control, and Dynamics*, vol. 13, no. 5, pp. 926–928, 1989.

72. S. More, and J. Singla, "Machine learning techniques with IoT in agriculture," *International Journal of Advanced Trends in Computer Science and Engineering*, vol. 8, no. 3, pp 742–747, 2019.

73. S. Mangrulkar, "Artificial neural systems," *ISA Transactions*, vol. 29, no. 1, pp. 5–7, 1990.

74. B. Kaur, H. Sadawarti, and J. Singla, "A neuro-Fuzzy based intelligent system for diagnosis of renal cancer," *International Journal of Scientific & Technology Research*, vol. 9, no. 1, 2020.

75. S. K. Yewale, and P. K. Bharne, "Hand gesture recognition using different algorithms based on artificial neural network," in *2011 International Conference on Emerging Trends in Networks and Computer Communications (ETNCC)*, Udaipur, India, pp. 287–292, 2011, doi: 10.1109/ETNCC.2011.6255906.

76. E. Stergiopoulou, and N. Ã. Papamarkos, "Engineering applications of artificial intelligence hand gesture recognition using a neural network shape fitting technique," *Engineering Applications of Artificial Intelligence*, vol. 22, no. 8, pp. 1141–1158, 2009.

77. J. Rege, A. Naikdalal, K. Nagar, and P. R. Karani, "Interpretation of Indian sign language through video streaming," *International Journal of Computer Sciences and Engineering*, vol. 3, no. 11, pp. 58–62, 2015.

78. V. Kecman, "Support vector machines – An introduction," *Support Vector Machines: Theory and Applications*, vol. 1, pp. 1–47, 2005.

79. W. Chen, and Z. Zhang, "Hand gesture recognition using sEMG signals based on support vector machine," in *2019 IEEE 8th Joint International Information Technology and Artificial Intelligence Conference (ITAIC)*, 2019.

80. K. Tripathi, N. Baranwal, and G. C. Nandi, "Continuous Indian sign languagegesture recognition and sentence formation," *Procedia – Procedia Computer Science*, vol. 54, pp. 523–531, 2015.

81. J. Singha, and K. Das, "Automatic Indian sign languagerecognition for continuous video sequence," *ADBU Journal of Engineering Technology (AJET)*, vol. 2, no. 1, pp. 117–126, 2015.

82. J. Singha, and K. Das, "Recognition of Indian sign language in live video," arXiv preprint arXiv:1306.1301, vol. 70, no. 19, pp. 17–22, 2013.

83. S. V. Gumaste, "Euclidean distance based classifier for recognition and generating Kannada text description from live sign language video," *International Journal of Recent Contributions from Engineering, Science & IT (iJES)*, vol. 5, no. 3, pp. 41–57, 2017.

84. G. Guo, H. Wang, D. Bell, Y. Bi, and K. Greer, "KNN model-based approach in classification," in *"OTM Confederated International Conferences" On the Move to Meaningful Internet Systems*, pp. 986–987, 2003.

85. A. Nandy, J. S. Prasad, S. Mondal, P. Chakraborty, and G. C. Nandi, "Recognition of isolated Indian sign languagegesture in real time 2 generation of ISL video training samples," in *International Conference on Business Administration and Information Processing*, pp. 102–107, 2010.

86. N. Yadav, "Noval approach of classification based Indian sign languagerecognition using transform features," in *2015 International Conference on Information Processing (ICIP)*, pp. 64–69, 2015.

87. R. Nogales, and M. Benalcázar, "Real-time hand gesture recognition using KNN-DTW and leap motion controller," *Information and Communication Technologies*, vol. 1, pp. 91–103, 2020.

88. W. Jiangqin, G. Wen, S. Yibo, L. Wei, and P. Bo, "A simple sign language recognition system based on data glove," in *ICSP '98. 1998 Fourth International Conference on Signal Processing (Cat. No.98TH8344)*, vol. 2, pp. 1257–1260, 1998, doi: 10.1109/ICOSP.1998.770847.

89. A. Z. Shukor, M. F. Miskon, M. H. Jamaluddin, A. Ibrahim, M. F. Asyraf, and M. Bazli, "A new data glove approach for Malaysian sign languagedetection," *Procedia - Procedia Computer Science*, vol. 76, pp. 60–67, 2015.

90. P. Lokhande, R. Prajapati, and S. Pansare, "Data gloves for sign language recognition system," *International Journal of Computer Applications*, vol. 1, pp. 234–248, 2015.

91. M. A. Ahmed, B. B. Zaidan, A. A. Zaidan, M. M. Salih, and M. M. Lakulu, "A review on systems-based Sensory gloves for sign language recognition state of the art between 2007 and 2017," *Sensors*, vol. 18, no. 7, p. 2208, 2018.

92. W. Li, P. Shi, and H. Yu, "Gesture recognition using surface electromyography and deep learning for prostheses hand: State-of-the-art, challenges, and future," *Frontiers in Neuroscience*, vol. 15, pp. 233–242, 2021.

93. E. Ceolini, C. Frenkel, S. B. Shrestha, G. Taverni, L. Khacef, M. Payvand, and E. Donati, "Hand-gesture recognition based on EMG and EVENT-BASED camera sensor fusion: A benchmark in neuromorphic computing," *Frontiers in Neuroscience*, vol. 14, pp. 347–356, 2020.

94. A. G. Jaramillo, and M. E. Benalcazar, "Real-time hand gesture recognition with EMG using machine learning," in *2017 IEEE Second Ecuador Technical Chapters Meeting (ETCM)*, 2017.

95. M. E. Benalcazar, J. González, A. Jaramillo-Yánez, C. E. Anchundia, P. Zambrano, and M. Segura, "A model for real-time hand gesture recognition using electromyography (EMG), covariances and feed-forward artificial neural networks," in *2020 IEEE ANDESCON*, pp. 1–6, 2020.

96. Q. Pu, S. Gupta, S. Gollakota, and S. Patel, "Whole-home gesture recognition using wireless signals," in *Proceedings of the 19th Annual International Conference on Mobile Computing &Networking - MobiCom '13*, 2013.

97. W. He, K. Wu, Y. Zou, and Z. Ming, "WiG: WiFi-based gesture recognition system,"in *2015 24th International Conference on Computer Communication and Networks (ICCCN)*, pp. 1–7, 2015, doi: 10.1109/ICCCN.2015.7288485.

98. H. Abdelnasser, M. Youssef, and K. A. Harras, "WiGest: A ubiquitous WiFi-based gesture recognition system," *2015 IEEE Conference on Computer Communications (INFOCOM)*, pp. 1472–1480, 2015, doi: 10.1109/ INFOCOM.2015.7218525.

99. D. Jiang, M. Li, and C. Xu, "Wigan: A wifi based gesture recognition system with gans," *Sensors*, vol. 20, no. 17, p. 4757, 2020.

100. D. A. Contreras Alejo, and F. J. Gallegos Funes, "Recognition of a single dynamic gesture with the segmentation technique HS-ab and principle components analysis (PCA)," *Entropy*, vol. 21, no. 11, pp. 11–14, 2019.

Index

For Product Safety Concerns and Information please contact our EU
representative GPSR@taylorandfrancis.com
Taylor & Francis Verlag GmbH, Kaufingerstraße 24, 80331 München, Germany